Y0-BRT-464

The magnificence of the Grand Harbour. Fort St Angelo in the centre with Valletta in the backround.

G. 12.00

MALTA
AN ISLAND REPUBLIC

DESIGN AND TEXT
ERIC GERADA-AZZOPARDI

PHOTOGRAPHY
CHRISTIAN ZUBER

EDITIONS DELROISSE

TABLE OF CONTENTS

	Page
At The Crossroads of Destiny	1
Malta and Her Dependencies	25
The Archipelago	27
The Geology of Malta	39
A Different Place	51
Population, Race and Language	59
The People	
History	
In The Beginning	95
Neolithic Malta	
The People of Ghar Dalam and Skorba	97
The Temple Builders of the Copper Age	
Zebbug and Mgarr	98
Ggantija and Hagar Qim	99
The Hal Saflieni Hypogeum	101
Mnajdra and Tarxien	102
Neolithic Malta and Her Decline	103
The Bronze Age	
Tarxien Cemetery People	
Borg-in-Nadur	105
Bahrija	
The Cart Tracks	
The Entrance of Malta into History	
Phoenicians and Carthaginians	127
Romans and Byzantines	132
The Arabs and the Normans	137
Medieval Malta	143
The Knights of St. John	
Outremer	151
Rhodes	155
The Golden Era of the Knights of Malta	
Prelude	158
The Great Siege of Malta	161
Valletta-City on the Hill	165
The Years of Consolidation	167
Decay and Decline and Fall	172
The Uprising against the French	193
The British Period	
Autocracy to Democracy and back to direct Rule	196
World War II	212
The Second Siege of Malta	
Malta After the War	217
Sources and Selected Bibliography	266 - 267
Map of the Maltese Islands	268 - 269
Acknowledgements	270

AT THE CROSSROADS OF DESTINY

The epic story of this cluster of little islands at the heart of the Mediterranean Sea, which now forms part of the Island Republic of Malta, has inevitably been turbulent and momentous.

It has been Malta's destiny to figure prominently in the long saga of Man on earth and to find herself closely involved in the evolution of civilisation within the Mediterranean basin. Always embroiled in the titanic struggles of this great sea, Malta lies at the crossroads of ancient, medieval and modern history. She has commanded an importance within the corridors of power of every given age quite disproportionate to her size, latent resources or physical condition. In this sense, Malta is perhaps unique.

The key to Malta's historic role is in her geographic location and in the magnificence of her natural harbours. Halfway between two great continents and two formidable oceans, Malta lies at the accurate centre of the Mediterranean. Located in the mainstream of cross-fertilisation of ideas and culture between the ancient East and the ascendant West, Malta has been in the vanguard of evolving humanity from the dawn of history. It was inevitable therefore, that she would be intimately exposed not only to the greater glories of Man and his achievements, but also to his vain struggles, conflicts and their attendant miseries.

Malta has suffered the calamities of war and gloried in many victories. She has enjoyed the relative calm of the prehistoric age and played host to the outstanding achievements of early stone age man. With the emergence of the early sea-faring trading communities, she derived great benefits. From then on, however, she fell prey to the vagaries of power and conflicts between the surrounding nations of the Mediterranean.

Often invaded, all too often ravaged and enslaved, Malta passed from one power to another; sometimes forgotten, often bartered and even sold; always sacrificed to the cause of the day. Her fortunes were always largely dictated by those of her occupiers. When fortunes rose, it was as a result of events which brought misfortune on others. A sudden prosperity would equally suddenly disappear only to be replaced by deprivation and sacrifice at the hands of the latest coloniser.

Malta's passionate adherence to a religious ethic is perhaps the only stable element present throughout her long history. Worship of the Prehistoric Venus, the sacred monuments of megalithic and neolithic man were to be followed by an early conversion to Christianity in AD 60. From the Roman days of St. Paul, Malta has clung tenaciously to her Christian heritage. This was the cause of much of her direct involvement in the death struggle between Christian and Moslem in the Middle ages. It was as a result of successfully withstanding the Siege of Malta in 1565, that the penetration of Moslem supremacy into Southern Europe was finally repulsed. The subsequent decline of Moslem power was no accident of history. It was during this Great Siege that, for the first time, the Maltese were party to an organised defence of this Island, as a brave ally of the Christian power of the day.

Nearly two hundred and forty years later, in 1800, they were again to organise themselves, to blockade the French in Valletta and finally, with the fortuitous assistance of the British, to bring to an end Napoleon's brief conquest of the Islands.

Another historic siege was to take place one hundred and forty two years later, at the height of the Second World War. Committed body and soul to the Allied cause against the Axis powers, the Maltese were again to display the formidable qualities of courage and tenacity, endurance and gallantry in Malta's last great battle as an impregnable island fortress. For their heroism, the Maltese were awarded the George Cross by King George VI.

It was the Second World War which led to de-colonisation and the gradual decline of the great empires of modern history. In the wake of that decline, Malta was destined to abandon her role as an island bastion in the power struggles of the Mediterranean. Finally, the Maltese were to have a say in their future destiny.

We have come full circle. Perhaps for the first time since the early peaceful days of neolithic man, Malta "regained" her sovereignty when she became constitutionally independent in 1964. She was declared a Republic in 1974.

On the 31st March 1979, the British military presence in Malta ended. From the historic bastions of the Grand Harbour, once the heart of British naval power in the Mediterranean, the Maltese bade farewell to an old friend. On this site, a monument depicting Britain's departure was erected as a fitting tribute to her long association with Malta. This monument also serves as an expression of the steadfast friendship of the people of the Island Republic of Malta.

The era of foreign military bases on Malta is now over. A peace economy based on industrial growth replaces it. In her new role, Malta now promotes an image of friendship with all neighbouring countries in the spirit of peace and co-operation.

The struggle for self reliance is not over. Malta is at the crossroads of destiny. Along the path of progress in peace, a new era unfolds.

« Guard O Lord, as always Thou hast guarded,
this sweet motherland whose name we bear,
clothed by Thee in radiance most fair :
Wisdom to her rulers, Almighty grant
to just employer and to worker, health :
Strengthen Malta in unity and peace.»

Dun Karm : Maltese National anthem [1]

1. Translated from the Maltese original. Dun Karm is Malta's national poet.

ĦELSIEN u SLIEM
MALTA 31-3-79
5c

MALTA
31-3-79
ĦELSIEN u SLIEM
7c

MALTA
31-3-79
ĦELSIEN u SLIEM
8c

REPUBBLIKA TA MALTA

31st March 1979 *The Union Jack is lowered.*

The Maltese flag is hoisted.

The hand of friendship.

From an island fortress to a centre of peace.

Page 10/11 Peace at Marsaxlokk.
12 The peace of the evening.
13 Peace in the countryside.

minn gżira-fortizza għal ċentru ta' paċi

spirotechnique

The modern House of Representatives in Valletta designed by the Maltese architect Richard England.

Malta has a democratic and republican form of Government. The Constitution provides for a uni-cameral legislature. The House of Representatives is the highest law-making body which together with the President of the Republic of Malta, constitutes Parliament.

The House is composed of 65 representatives who are elected by the people by universal franchise. The election is based upon the principle of proportional representation by means of a single transferable vote from thirteen electoral divisions in the Islands. The life of Parliament is for five years, unless dissolved sooner.

Auberge de Castille et Leon in Valletta. Built by Grand Master Pinto and designed by the Maltese architect Domenico Cachia. A fine example of 18th Century Maltese Baroque. The auberge is now the Office of the Prime Minister. The monument in the front is to Manwel Dimech, an early Maltese Socialist.

MALTA AND HER DEPENDENCIES

The Archipelago

The Maltese Islands are small. Malta is the largest and most important of the group which lie at the heart of the Mediterranean Sea.

Malta has a total area of 95 square miles. Her name is derived from the Phoenician «malet» meaning refuge: a clear reference to the many safe harbours which the early Phoenicians found in Malta. The Greeks called her «melitte» which became «melita» to the Romans. It is possible that this is a reference to the excellence of Maltese honey which was produced even then. It could also be a more hellenised and corrupt form of the original «malet».

Gozo, the most northerly Island of the Archipelago has an area of only 26 square miles. To trace the name of Gozo, one must first take its Maltese form which is «Ghawdex». Because the shape of the Island is that of a lentil, it is likely that the Phoenicians called it «Gul» or «Gawl» or even «Gaulos». The Greeks, and later the Romans, turned this into «Gaudos». The Arabs found this difficult to pronounce and called it «Ghawdex». Soon the Spanish found that name impossible for them and settled for «Godzo» and finally «Gozzo».

Between Malta and Gozo, in the Fliegu channel is the small Island of Comino with its own smaller sister Island, Cominotto; together these two islands hardly exceed one square mile. Comino owes its name to the abundant supply of cummin seed; an aromatic herb that the local community once produced for flavouring many a dish.

There are also the legendary islands. Firstly, the Island of Filfla which lies about four miles to the south east of Malta facing the prehistoric temples of Mnajdra and Hagar Qim. It is sufficiently distant to have been used as a gunnery target in the recent past. Today, peace prevails; Filfla is a bird sanctuary and an important breeding ground for the stormy petrel or «Kangu ta' Filfla». It also appears to be the traditional home of an ancient colourful lizard, dark green with white spots, which is unique to the Island. The name Filfla is derived from filfel, a pepper which grew profusely before the bombs and shells obliterated all vestige of it.

The other two Islands of legend are known simply as «Il-Gzejjer» or «The Islands». They stand at the entrance to St. Paul's Bay. It is here that St. Paul is reputed to have been shipwrecked in AD 60, whilst on his way to Rome; an event recorded in some detail in the Acts of the Apostles by St. Luke.

To complete the archipelago, there is finally a mere rock of an islet, lying in Dwejra Bay on the west coast of Gozo. The steep and massive 'Fungus rock' is also known as 'Il-Gebla tal-General' or 'The General's rock'. It was once believed that the fungus on the rock was effective against dysentery and haemorrhages. It is also said that, in the middle ages, an Italian general in charge of quarrying on it, fell off its steep side and was drowned within arm's length of the rock. He has never been forgotten.

The combined area of the Maltese Islands is nearly 122 square miles, roughly the same area as the Isle of Wight. The size of Gozo is nearly equal to that of Hong Kong Island or that of Bermuda in the Caribbean. Malta is slightly larger than the total area of the Channel Islands.

Distances are obviously short on the Islands. The length of Malta's coastline is 85 miles, while that of Gozo is a mere 27 miles. The maximum distance in Malta from North West to South East is about 17 miles. The greatest width is only 9 miles. Gozo is 9 miles long and hardly 5 miles wide.

← *Page 16/17 Gnejna Bay, Malta.*
18 Multi-coloured boats at Marsaxlokk.
19 Rural scene in Gozo.
22 St Julians Bay, Malta.

THE GEOLOGY OF MALTA

The Maltese Islands have not been blessed with an abundance of natural resources. There are no mountains or rivers or forests. No traces of mineral or metal deposits have ever been found. There is no coal or oil or precious stones. Malta has to struggle for her water and devise ingenious ways to collect and store the little rain that falls. Malta is still four-fifths rural. The tenacity and ingenuity of the Maltese farmer keeps the sparse arable land that exists in its present intensive state of cultivation.

Ask the average Maltese about Malta's resources. After a spontaneous monologue in praise of the Maltese and their great history, he will then pause and suggest the sun, the sea and the local stone - limestone. All these are clearly plentiful.

The Islands were formed during the oligocene and miocene periods of the tertiary epoch. Traces of quaternary deposits have largely eroded with time. The basic composition is limestone: Upper coralline, a layer of golden globigerina and lower coralline. There is then a topping of blue clay and a little greensand. The coralline limestone is responsible for the presence in Malta and Gozo of the megalithic slabs of the stone age people. The soft globigerina, easy to quarry and to carve, is the traditional building material found everywhere. Blue clay is highly fertile and because there is more of it in Gozo, that Island is more fertile.

The Island of Malta consists of a low, faulted plateau which tilts from the cliffs to the west to the bays of the north and east. The indentations in the coastline are an interesting feature. In the north, the fault lines account for both the barren rocky ridges and the alluvial fertile valleys following the hills. Gozo too, lies slightly tilted and slopes into the sea from her south western side.

The predominance of the limestone everywhere is most striking. Combined with the heavy sun of the torrid Maltese summer and the enveloping seascape surrounding the Islands, the contrast is vivid and even magnificent.

Here, Man has successfully fused these component elements ot form a total landscape which is at once structured and etherial; but never contrived.

Fort Chambray on the Island of Gozo. One of the last of the fortifications built by the Knights of St John in 1749. It is now a mental hospital.

Page 26: 'Il-Gzejjer' also known as St Paul's Islands. Traditionally, the site where St. Paul was shipwrecked in A.D. 60 whilst on his way to Roma.

Page 27: The Island of Filfla with Malta about four miles in the distance. It is now a bird sanctuary and an important breeding ground for the stormy petrel.

Pages 28-29: The Islands of Comino and Cominotto, with Malta in the haze.

A look-out tower at Fomm-ir-rih overlooking Gnejna Bay and Ghajn Tuffieha Bay. There are a number o similar towers dotted round vantage points on the coasts. The towers formed part of an alarm security syster when raiding fleets were sighted.

The limestone quarry.

Wied Iz-Zurrieq, Malta.

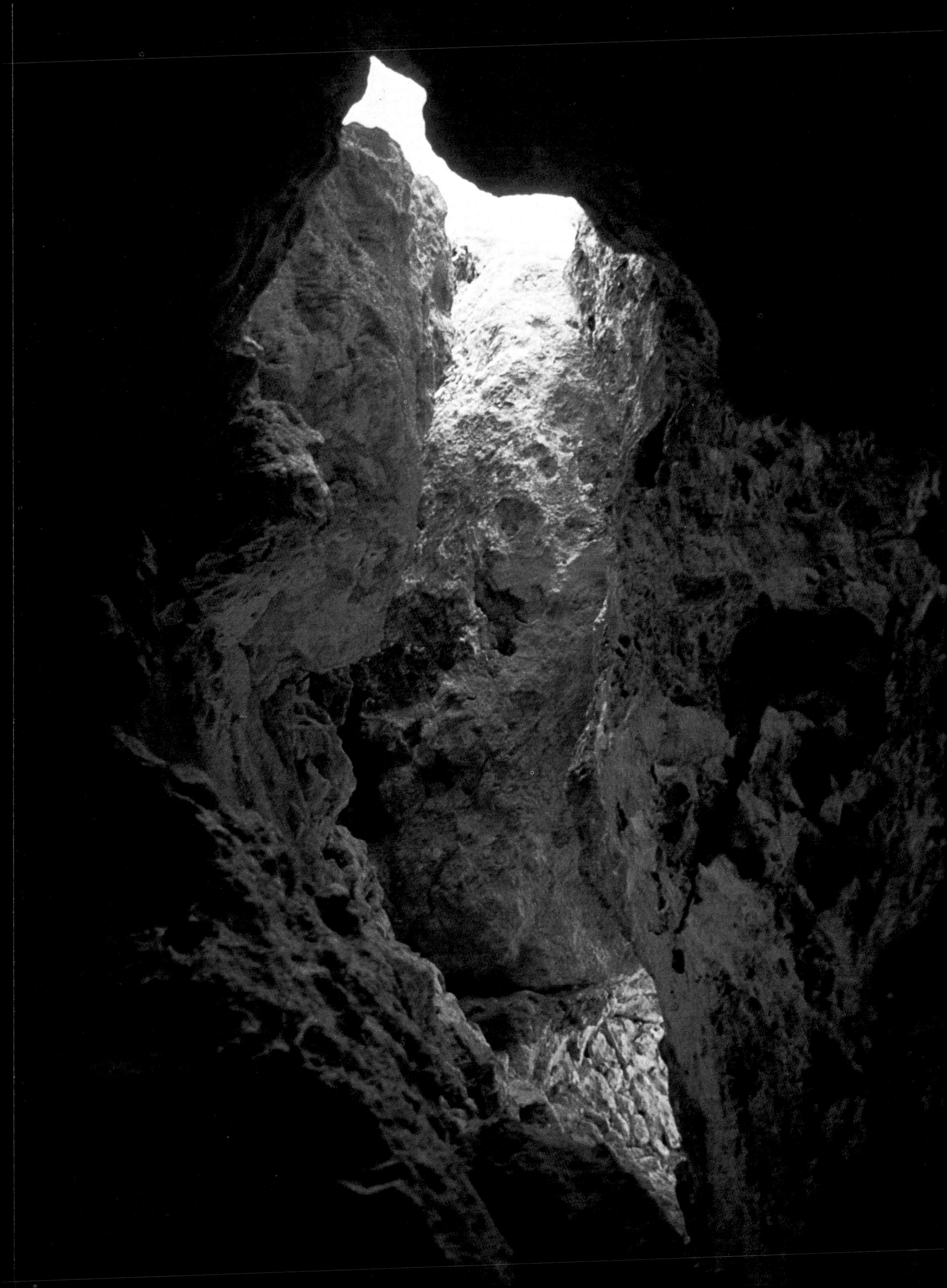

Dried limestone and drying garlic.

'alypso's Cave in Gozo.

Maturing grapes.

Between the hills lie the sweeping fertile valley

A DIFFERENT PLACE

Quite unlike anywhere else in the Mediterranean littoral, Malta expresses an individuality in her Island landscape which is not to be found in the other islands of the Mediterranean. The essence of this golden landscape stems from the magnificence of its stone. The limestone of the more mature mountains of Spain or those of North Africa is quite unlike that of the rolling hills of Malta. The sombre mountains of the European coasts are also different. Although Malta is very Mediterranean in appearance, the first impression is always one of surprise ; because her landscape is subtly different.

These stone Islands emerge abruptly and steeply from the surrounding sea, as if to punctate their very presence. From the air, the geological faults are clearly noticeable as they lead to the many indentations of the coastline. To the south west, the sheer cliffs at Dingli rise sharply and swiftly. Here the coastline is inhospitable. The landscape is wild and dramatic. More gentle are the slopes leading to the east and north and stretching into the sheltered bays and creeks. It is here that the important natural harbours kiss the stone bastions of Valletta and embrace the old cities and their extensions - the new towns of modern suburbia.

Shape and form prevail in Malta. There is surprise in the geometric power of the Maltese landscape. The activity of the stark Mediterranean light on form brings out the intrinsic character of the Islands. Malta is a land of stone; and this shapes everything. It is almost like a conspiracy. There is a near faultless relationship of scale between Man, his stone buildings and walls and the virgin land mass. Man-made form is cubic and the pattern is rectangular. The colour is that of limestone. Depending on the light, its shades may be pollen yellow, saffron or ochre and sometimes, gold.

Mile upon mile of flat roofed houses, tightly packed and parcelled within ancient, fortified walls or clustered into tiny villages, each dominated by modelled soft domes and elegant steeples of Baroque parish churches; the simple and the elaborate combine. Then, dotted around the coasts are the massive stone forts and the look-out towers tom-peeping over the bays. Here and there, amidst a patch of green, a castle may be seen.

Mile upon mile of flat roofed houses, a patch of green here and there and Mosta Church with its magni-cent dome.

Every patch of arable land is carefully nursed. The Gozo countryside with freestone terracing at every level, stretching endlessly.

And then again, the long lines of heavy defence of the citadels appear in the silent form of bastions and bulwarks, ramparts and redoubts, cavaliers, curtains and counterscarps protecting the great houses, the palaces and the piazzas. Beyond the old cities but extending from them lies the outcrop of modern urban spread; cube upon cube, more houses, cheek by jowl, more stone.

The intermittent countryside appears with a spontaneous simplicity so characteristic of rural Malta. The whole area appears to be allocated to rectangular plots with free stone terracing at every level, stretching endlessly. It is clear that every patch of arable land is carefully nursed. Between the terraced hills lie the sweeping fertile valleys, equally patterned, and obviously benefitting from both the ancient and more modern irrigation systems. Water is vital to this arid landscape, if it is to maintain its productive momentum and agricultural promise. No wonder then, that the collection of rainwater is taken so seriously in the Maltese islands.

The Island landscape of sky, sea and stone does not present a conventional image of scenic beauty. What emerges is a reflection of the successful collaboration between Man and nature. The result is a heavy presence of earthy mysticism, which has freely penetrated into the life patterns and culture of the Islanders.

Today, there is much hustle and bustle in modern Malta; but the mysticism remains. The giant solid monoliths of the prehistoric temples bear testimony to the ancient mysteries, forever locked in their walls of worship. In this splendour, they stand erect and inert in the abyss of silence.

The 'Gran Castello' - Rabat, Gozo.

WELCOME TO GOZO

The fishing village at Mgarr harbour in Gozo. A maze of colour.

Page 42: Weathered bastions guard their secret history.

Page 43: Silent gun on Gozo.

Pages 44-45: The south-west coast of Malta. The stone Island emerges abruptly and steeply out of the blue.

Fishing in the blue Mediterranean - Mgarr ix-xini, Gozo.

UP TO DATE GARAGE

POPULATION, RACE AND LANGUAGE

There are about 310,000 Maltese living on the inhabited Islands of Malta, Gozo and Comino ; although the latter only supports a token community of about 50 persons. There are as many Maltese living abroad as there are in Malta. The Maltese took to migration from the earliest times and made homes in many parts of the Mediterranean especially in North Africa and Egypt. In modern times they have settled mainly in Australia, North America and the United Kingdom.

There has been a rapid increase in the population of the Maltese Islands over the last one hundred years or so, but this has now levelled off. Indeed, the present birth rate is one of the lowest in Europe. The population density of 2,500 persons to the square mile is one of the highest. The fact that the population is concentrated into relatively few urban areas considerably aggravates the problems of density. However, it is this that has saved the rural character which the Maltese Islands still enjoy to this day.

The origins of the Maltese are the subject of much argument and controversy. If one discounts the sparse colonisation by neolithic and bronze age man, who arrived from Sicily and mainland Italy, the likelihood is that the Maltese are of early Caucasian stock. They stemmed from the Phoenicians who were moving westwards into most parts of the Mediterranean from the Levant and the head lands of the Nile. These early settlers were clearly subject to many external influences that were shaping the whole region. The melting pot had commenced. When the Greeks came, they co-existed well with the Phoenicians. It was when the Carthaginians arrived after them and entered into a death struggle with the Romans during the Punic Wars that this Maltese mixture was brought into contact with the new Latin civilisation. For the next seven centuries, the Romans dominated the Islands.

The final division of the Roman empire brought the influence of Byzantium to Malta, which in time gave way to the period of Arab rule lasting for two hundred years. The Maltese were thus again brought into contact with their early semetic origins. The Europeanisation of Malta can be said to have started with the arrival of the Normans from Sicily in the late eleventh century.

While retaining much of their Armenoid origins, the Maltese therefore intermingled with the whole spectrum of peoples who then dominated the Mediterranean sea. The modern hybrid is therefore an interesting mixture which is primarily Levantine with a shade of Graeco-Roman and Arab, tinged with many parts European, mainly Sicilian, Neopolitan and Spanish. However, a further influx of European blood was assimilated during the Middle ages ; mainly Italian, Spanish and even French. The long period of British rule has also left traces of the Anglo-Saxon.

These racial influences clearly had their effect on the language which the modern Maltese speaks today . Maltese is now a language in its own right and spoken throughout the Maltese Islands. Its origins are also 'Punic' which is basically a semetic off-shoot of Phoenician. The language remained in use through the long periods of Carthaginian and even Roman rule. When the Arabs arrived in AD 870, it gave way fairly easily and comfortably to the onrush of Arabic words. The Semetic base was thus further strengthened and even consolidated. This element of Maltese grammar has persisted to this day, despite the prominence of the Romance languages ushered in with the Europeanisation of Malta. From then on, the vocabulary became laced with words loaned from Sicilian, Italian, Spanish, French and latterly, even English. The Maltese alphabet is a successful transliteration of Semetic sounds but, uniquely, written in Roman characters.

Although widely spoken from the earliest times, the story of Maltese as a language is a sad one. Never recognised as a language of culture until recently, it was Italian which held sway certainly from about the nineteenth century onwards. Italian was the official language during the period of the Knights of St. John. During the British period, English and Italian became the official languages. It was not until 1934, after a long and bitter struggle which had political undertones, that Italian was withdrawn and substituted by Maltese.

Together with English, the Maltese language finally became the accepted official one to be used in the public administration of the Islands. Maltese also became the official language of the law courts for the first time.

As a direct result of the great assimilations and contact with a variety of languages, the Maltese have acquired a natural linguistic aptitude. Today, they are generally tri-lingual and fluent in English and Italian in addition to their native Maltese.

But this, Friend of my heart,
I have; this God gave me,
this word in whose beauty
you see me now delighting.
Who said that it is heavy, and crawls
like a snake creeping on its belly ?
See, how light it is,
How suddenly it leaps and soars;

And in the word that rises
from my heart to my mouth
pure, sweet and beautiful
even as my mother taught me,
As once in Valletta, as you still
today hear it in the villages
without trappings on its breast,
poor, but possessing all.

See how lovely it is
like a bunch of May flowers:
not sweeter is the fragrance
of basil and pennyroyal.

A. Cuschieri: The Maltese Language [1]

1. A.J. ARBERRY A Maltese Anthology (Oxford University Press, London 1960) p. 223-226

The imposing Cathedral at Rabat in Gozo with its restrained baroque facade designed by the Maltese architect Lorenzo Gafá and built in 1711. To the left of the Cathedral in the charming square is the old Bishop's Palace built in 1620. It now houses the Law Courts and the public registry.

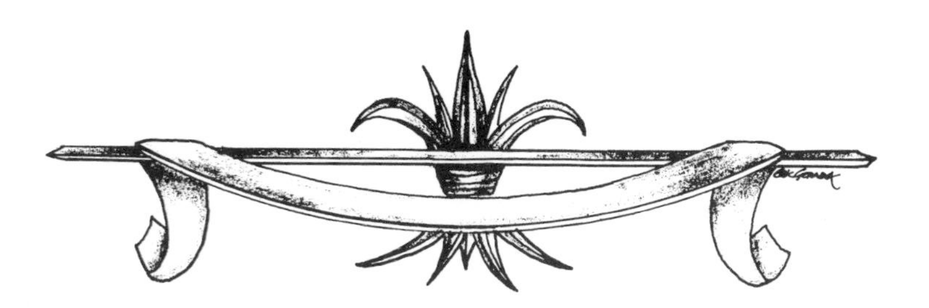

Festa day at the Parish Church of St Paul in Rabat, Malta. Founded in 1575, the Church also houses the Sanctuary of St Publius - the first Bishop of Malta. A ticker tape welcome and a few fireworks for the arrival of the statue of the Patron Saint.

Page 56: The interior of Gzira Parish Church.

Page 57: The market stalls outside Floriana on a Sunday morning.

Page 58: The Valletta market stalls. Nuns and tourists strike a bargain.

THE PEOPLE

Like most Islanders everywhere, the Maltese are highly individual. Their separateness, combined with a history of siege and occupation has inevitably produced a character which is at once proud, stubborn and insular. At the same time, the Maltese are overwhelmingly friendly and accomodating: their hospitality is legendary. There is a subtlety here which can be disarming. For a start, the pride of the Maltese is not easily discernible. This is because of a prevailing sense of prudence and humility which underlines the Maltese way of life. Only if abused, will this inherent pride be displayed. The strong religious beliefs and family loyalties of the Maltese must be respected. They will show much flexibility in their dealings; but there will always be an iron resolve in matters affecting their progress from which they are unlikely to waiver.

The Maltese are notoriously cautious. Their experience of frugal times also makes them thrifty. They will always 'save for a rainy day'. Yet they love a gamble now and then. The Maltese will speculate and invest: their business acumen is well known in the Mediterranean. Despite the inbred caution, there is a clear adventurous streak in the Maltese. At one time, Malta produced the ablest corsairs in the Mediterranean. The Maltese took to the sea from the earliest times and they founded many trading communities dotted around the North African coasts. In recent times, they have emigrated to Australia and Canada and proved to be exemplary settlers.

The Maltese are hardy and noted for their stamina and perserverance. They have a capacity for hard work and long hours which is curiously un-Mediterranean. Given a good cause and strong leadership, the Maltese will endure much hardship and sacrifice. Their history is ample proof of their moral fibre.

The concept that true happiness lies in moderation is still much cherished by the Maltese. There is also a 'no nonsense' approach to life and decision-taking which may appear to be out of tune with the easy-going pace of life. The 'take it or leave it' principle is freely applied here. At the same time, this strict code is tempered with a magnanimity which can be most perplexing. The generosity of the Maltese is very beguiling. They will go to great lengths to save pennies, but they are then prepared to spend pounds.

The Maltese are rumbustious, gregarious and love a good party. They take their revelry very seriously. Their carnivals and folk religious Fiestas, which they call Festas, are always highly organised, colourful, noisy and hugely enjoyable. The Maltese love the hilarious and they have a special feeling for the outright ridiculous. The basis of this earthy humour is farce.

The Maltese respect for fun is never seriously abused and they are hardly ever extravagant - except in their passion for fireworks! Moderation will then suddenly fail them. With every bang and burst, the Maltese will shed their prudence, their sobriety and even their caution. Soon, there is a joyous ecstasy and abandon which only noise, movement, contrast and colour can stir in the Maltese. The firework display is the final embodiment of the fun of the Festa. This is the stuff of a rich and varied folklore ; an amalgam of ancient pagan rites to ward off evil spirits, fused to Christian rituals and embracing the natural flamboyance of the Southern European.

The Festa is also an outlet for the equally compelling passion of the Maltese for ceremonial. Ceremony is always an expression of the dignity of an occasion; and dignity strikes a respectable chord in the Maltese personality. To give added dignity to the grand occasion, what better way than to combine this with a religious feast. Hence, the important ingredients of the successful Festa. First, the dignity, solemnity and reverence of the religious procession. This will give way to the more earthly exultations of church bells and village bands. Then to end it all with a bang, comes the parade of fireworks, exploding and reverberating everywhere and lighting up the heavily decorated parish square. The old, the young and the very young, candifloss in hand and ice-cream dripping over 'Sunday *bests*' will all cheer and clap with every flash, as they gaze upwards towards the heavens, now filled with the puff smoke of Roman candles and rockets, bombards and petards. This is the height of revelry and festivity; tomorrow is another day.

Religion is important to the Maltese. They are predominently Roman Catholic. Their Christianity is one of the oldest in Europe. Although the great power and influence which the Church once wielded in the parishes is slowly being toned down, the Maltese still enjoy the benign security of a deep faith. This they practice with an ardour which is most touching and even impressive. Belonging to this great Church is a source of infinite pride to the Maltese and they wear their Roman Catholicism like a second nationality, proudly and sometimes fiercely.

The Maltese then, are a happy and a temperate people. Their heroism through siege and sacrifice has only strengthened their resolve in the pursuit of peace and progress. Idealists they may be, but because they have known hard times and deprivation, their singular individualism makes them restive and even frustrated. Having been denied a voice in their own affairs or those of their region for so long, it is natural that they should now be more resolute in the search for solutions which would ensure not only their progress, but also bring about the kind of peace and co-operation which the Mediterranean has never known.

As an Island people they are a microcosm of the history of this great sea. They are therefore well steeped in the Mediterranean experience. It is their desire to draw upon the lessons of history and attempt the overdue reconciliation between the peoples to the north and to the south of the Mediterranean. This, they can do if they are seen to be the trusted friend of both.

And the crowds stood up as one man and cried. 'I am Maltese !
Woe to him who despises me, woe to him who decieves me! '
And the crowds sang as one man, and made heard on the winds
the Anthem of our Malta; and the voice was victorious
Over the slumber of the cowardly past, the slumber of apathy
when our soul was lying in a foreign bed,
And the ghost of Vassalli arose from the depths of the tomb
and cried aloud; 'Now at last I shall find my rest'

Ruzar Briffa: The Anthem and the crowd. [1]

1. Arberry: Op. cit. p. 239
This poem was written in 1945 to commemorate a protest made by the crowd when the Maltese National Anthem was not played along with the Yugoslav and the British at a football match between Malta and Hajduks.
Vassalli was an early Maltese patriot.

A glimpse at the people.

Page 63: 'Tal-Haxix' - at the 'greengrocers'.

Page 64: The very young.

Page 65: The young.

Page 66: Pride of home with a bit of do-it-yourself ingenuity.

← *Page 67: Churchill's generation and the not so young.*

"You name it. I've got it!".

Pages 71-72: Up the village path on a lazy afternoon.

Tombola and a little gossip.

Forever young.

Doors and door-knobs - always spotless!

Page 76: Silver filigree at its best.

Page 77: Malta glassware. →

Folklore cabaret on an evening out.

oving with the times - fishing nets and colourful carrier bags.

A thriving industry - suits of armour for the 20th Century tourist.

The Maltese Islands have a long tradition in lace-making. Gozo lace is particularly fine. Handed down from generation to generation, the skill and craftsmanship of the lace-maker is as high as ever.

Three generations of lace-makers - grandmother, daughter and grandchild.

1921
1971
MADONNA BERIKNA

The Mosta 'fireworks' brigade - and the fireworks by night.

'esta day at Gzira Parish Church.

TAXI
STAND
FOR 2
EVERMOND
COSMETICS
79671

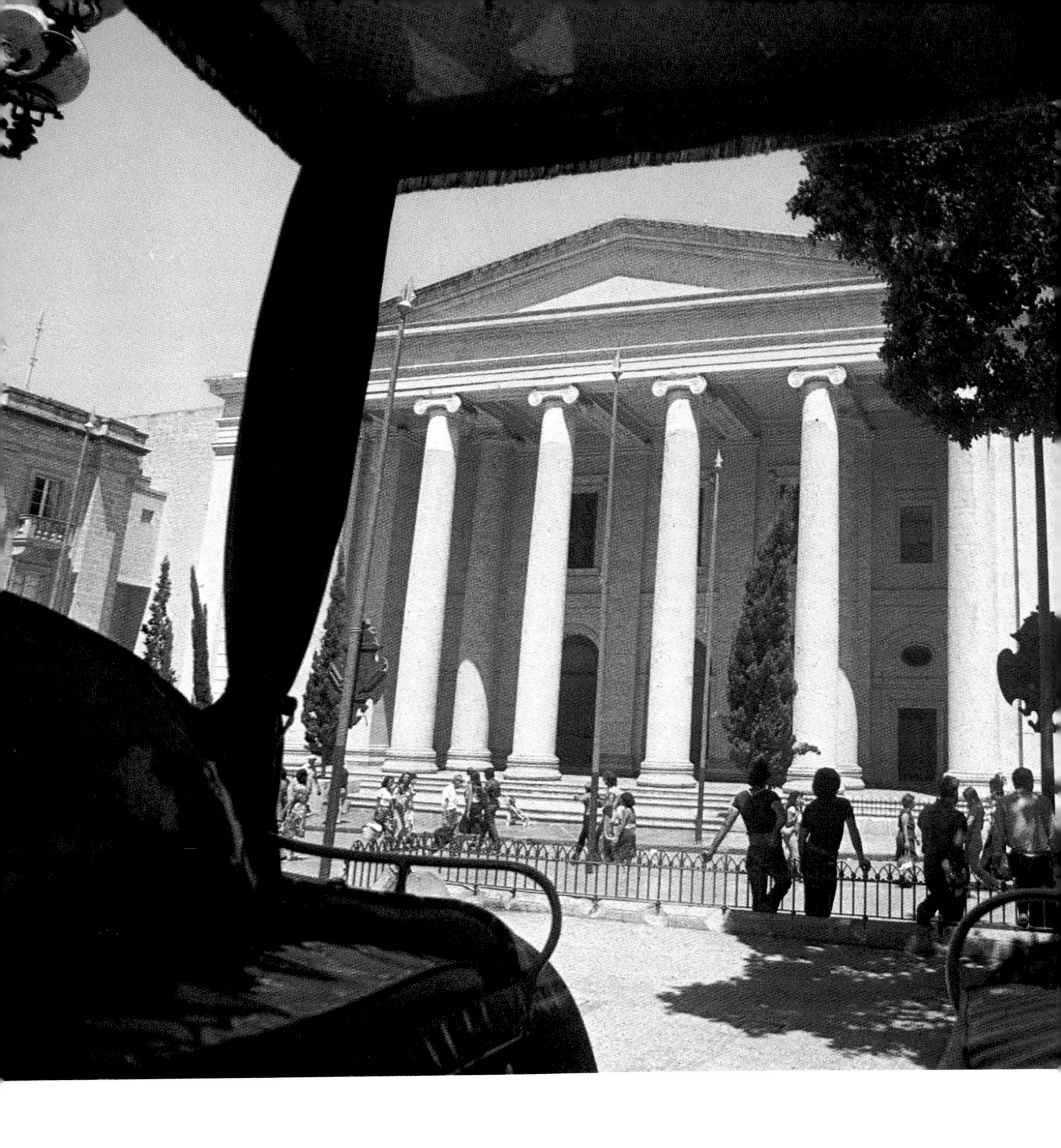

Page 86: Gaily decorated street in preparation for the festa.

Page 87: Queen Victoria statue in Valletta. It was unveiled in 1891 to commemorate the Queen's Jubilee four years earlier. In white parble, the statue is the work of the Italian sculptor Guiseppe Valenti of Palermo.

The newly built Law Courts in Valletta.

Mdina - the 'silent' city.

The 'Lion rampant' from the armorial emblems of Grand Master Manoel de Vilhena.

Pages 92-93: The fortified ancient city of Mdina. Its noble and glorious past is well preserved. The capital of Malta for many years, it was known as Melita in Roman times. In 1427, Alfonso V of Aragon named it Notabile. After Valletta was built in the 16th Century, Mdina ceased to be the Island's capital and became known as Cittá Vecchia. Today, the city retains its medieval character with a dignified silence which is jealously guarded. ➡

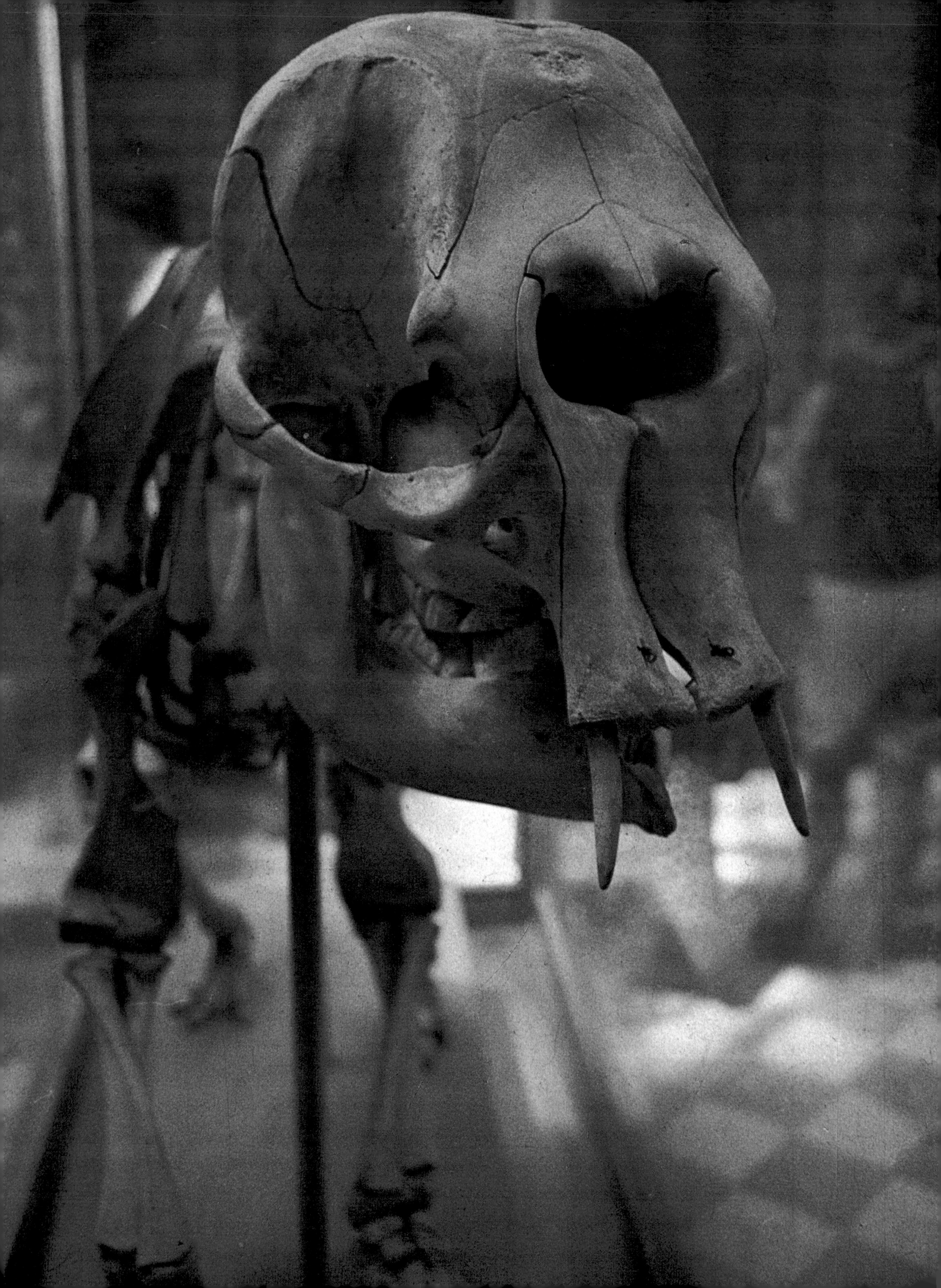

HISTORY

IN THE BEGINNING

In the beginning, there was Tethys. To the geologists, Tethys was the immense ocean which covered the whole of the modern Mediterranean. In Greek mythology, Tethys was the daughter of Uranus, the sky, and of Gaea, the Earth. She was married to her brother Oceanus, the great ocean god of the entire known world. Together, they had three thousand sons; the rivers, and three thousand daughters; the water nymphs, along with Metis (Wisdom), Tyche (Fortune) and Styx (the infernal river).

With the crumpling of the Earth's crust as a result of volcanic activity, Tethys started to recede. Two great lakes were formed on either side of a long central ridge of land, the highest point of which was the volcanic mountain of Etna in Sicily. From here, the ridge stretched South, past the highlands and plateaux of present day Malta and Gozo, and into the lowland plains below, extending to North Africa.

To the East of this landbridge, lay the waters of the eastern lake extending to Egypt and Syria. To the far west, the western lake was bounded by another landbridge connecting Africa to Spain.

Then the cataclysm came. The western landbridge, linking the Rock of Gibraltar and that of Ceuta in Morocco, was broken through, and the ocean roared in. The western lake was flooded. The central landbridge was over-run, as the gushing waters rushed through and finally united with the waters of the eastern lake. Only the highlands survived the floods. Marooned, there appeared for the first time the islands of the central Mediterranean in a unified great sea - the new Mediterranean.

That the Maltese Islands were joined to the landmass of Europe and Africa can be concluded from their geological formations.[1] During the fourth and last ice age, about a million years ago, these areas must have enjoyed a mild and temperate climate. From the highlands of Malta, could be seen the endless acres of land stretching south into North Africa. These plains were fertile, lush and green. Hippos and elephants from the south mingled with red deer, wild boar and antelope from the north. The roaming animals from Europe and Africa occupied this haven of peace and pasture until it was so rudely disrupted by the great floods about seven hundred thousand years ago. Then the waters came, and gradually cut off the Maltese islands from Sicily and North Africa. The isolation of today began.

The fossil remains of a variety of animals which survived this submergence were discovered in the Cave of Ghar Dalam - the Cave of Darkness, at Birzebugia and at Bur Meghez Cave near Mqabba in Malta. True, the skeletons found were those of dwarf elephants. This was the subject of much speculation and controversy. The likely explanation for their reduced size is that they were the degenerate descendents of animals of normal size so trapped on the newly formed islands. These elephants decreased in size with every new generation due to the scarcity of suitable food and the worsening conditions of their new environment.

There is also some controversy about the possible existence of Neanderthal Man on the Maltese Islands during the Palaeolithic age. Despite evidence that the Islands were well wooded, fertile and abounding in game, the likelihood is that they lay uninhabited for many thousands of years. They became a peaceful sanctuary for birds and a paradise for wild animals, free of the dangers of Man, the hunter. Indeed, Man reached the Maltese islands relatively late in time.

1. This is still a debateable point. That Malta was joined to Sicily seems certain. Some maintain that Malta was never joined to North Africa.

Mounted skeleton of a young elephant to show the approximate size of extinct species reputed to have been found at Ghar Dalam.

The rigours of the last ice age were gradually passing away from Europe. Changes in both flora and fauna were also taking place. The great hunting societies of Palaeolithic Man in Europe and elsewhere were already on the decline. To the east, hunting was giving way to farming. The domestication of plants and animals was leading to a more settled existence of family groups. Soon, early farming communities breeding cattle, pigs and sheep and cultivating wheat and barley were to make their appearance on the European and African coasts of the young Mediterranean. As land lost its fertility through cultivation, there was a continual migration of peoples and families from one region to the next. Man was on the move.

PREHISTORIC CHRONOLOGY [1]

Period	Phase name*	Approx date
Ghar Dalam Neolithic	Grey Skorba Red Skorba	5,000 - 3,750 BC
Copper Age	Zebbug Mgarr Ggantija Saflieni Tarxien	3,750 - 2,200 BC
Bronze age	Tarxien Cemetery Borg in-Nadur Bahrija	2,000 - 1,450 BC 1,450 - 800 BC 900 - 800 BC

* adopted from'type site'

1. The above prehistoric chronology is reproduced from the essay entitled 'Malta in Antiquity' by Professor *J.D. Evans* in Blue Guide Malta edited by *Peter McGregor Eadie* (Ernest Benn Ltd. London 1979). p. 30

Professor J.D. Evans of the Institute of Archeology in the University of London is a leading authority on Malta's antiquities.

'The dating of the prehistoric period in Malta before the beginning of the middle phase of the Bronze Age (Borg in-Nadur culture) is based almost entirely on a number of radiocarbon determinations made on material found in various sites. '

Quoted from J.D. Evans, Blue Guide Malta edited by Peter McGregor Eadie Ibid. p. 9.

NEOLITHIC MALTA

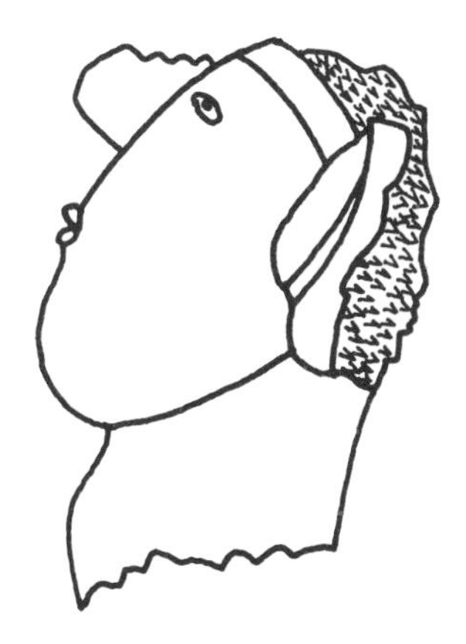

THE PEOPLE OF GHAR DALAM AND SKORBA

The colonization of the Maltese Islands came late; but it was an inevitable development. Malta is separated from Sicily by a mere 60 miles of sea. On a clear day, the Islands can be seen in a distant haze. From Sicily then, Man crossed to Malta either in coracles, small wicker framed boats covered in hide, or in plain wooden canoes. He found in Malta a fertile, virgin land, well suited to his pursuits of husbandry, hunting and a little fishing.

Evidence of New Stone Age Man, commonly labelled as 'Neolithic Man' was ascertained when impressed pottery dating to this period was found at Ghar Dalam, near Birzebugia in Malta. From the earliest times, this 'Cave of Darkness' seems to have been a hive of much activity. Not only were the bones of extinct animals of the Pleistocene age found deposited here, but a considerable amount of man-made pottery, identified with the Ghar Dalam people and those of subsequent phases, were also discovered. The pottery covers a wide historical span of time. It includes items belonging to the Grey and Red Skorba phases which followed that of Ghar Dalam proper. Pottery associated with the later Copper and Bronze age periods were also found at Ghar Dalam.

Certainly, the abundance of many natural caves in Malta and Gozo must have been of great benefit to the early colonizers. The caves offered secure and well protected dwellings for early Man and his animals; cattle, sheep, goats and pigs. In Malta, the cave of Ghar Dalam stands out as the most important then in use. There is evidence that, in Gozo, a series of about twenty caves at Ghajn Abdul were also inhabited.

The Ghar Dalam people developed new techniques in food production. They cultivated the fertile fields, especially those in the vicinity of the caves. It is likely that they grew club wheat, barley and possibly, even lentils. Besides making pottery, they chipped stone implements and axes. The use of metals remained curiously unknown to them. Soon, they were to erect simple huts and group them into tiny primitive villages. There is evidence of such huts at Skorba, on the outskirts of the village of Zebbieh in Malta. The Skorba people, as they are now referred to, farmed, hunted and fished in peace for close on 600 years. Because of variations in their pottery, this period is now generally divided into the Grey and the later Red Skorba phases. The pottery is almost identical to the Stentinello type belonging to the earliest agricultural communities of Sicily. From Sicily too, the new colonizers brought with them the religious fertility cult based on the female maternity goddess - the source of all life. Because of some of the contents found in the huts of the Skorba people, it can be concluded that religious ceremonies were conducted there.

Skorba was the first neolithic village to be found in Malta. In it were also discovered items of different prehistoric phases spanning that from Ghar Dalam to the much later Tarxien period. Indeed, the remains of a trefoil temple belonging to the Ggantija phase and later modified during the Tarxien period, can still be seen.

THE TEMPLE BUILDERS OF THE COPPER AGE

ZEBBUG AND MGARR

A further influx of immigrants from Sicily in about 3750 BC introduced a more advanced civilization which was to hold sway for the next thousand years. It ushered in a great cultural revolution which was to produce the world's most impressive prehistoric monuments in the form of the great stone temples of Malta and Gozo. Some of the finest examples of 'Megalithic' architecture (from the Greek 'Megas' meaning Big and 'Lithos' for Stone) can be seen in the Maltese Islands.

These early 'Maltese' were clearly starting to express an individuality unique to the Islands which was quite seperate, in its development, from that of neighbouring Sicily or other parts of the Mediterranean. The achievements of Man during this hectic and creative period were largely influenced by the now rampant religious cults and rituals in veneration of the fertility goddess but also linked to ancestor worship.

Clearly, the ample presence of easily cut limestone in Malta must have accelerated the advance and evolution of the early underground tombs which were cut into the living rock, as at Zebbug and Mgarr in Malta. Apart from the human and other animal bones, the Zebbug tombs yielded the most important discovery in the form of a small stylised statuehead, carved into a piece of globigerina limestone. It is at Mgarr that the oldest of the existing megalithic monuments was found. Here we are introduced to the first monumental facade enclosing a semi-circular forecourt. The oval building, over an area of about 900 square feet, is very much the proto-type for all the later temple complexes to be built in the Maltese Islands. Although no human remains were ever recovered from Mgarr, it is possible that the temples were used as tombs at first. The cult of the dead and its associated rituals must have certainly prompted a requirement for surface temples to be erected and thus satisfy the growing need to celebrate more properly the religious rites of the day. Soon, all the later surface temples were built as shrines for that purpose. They were to acquire a marked degree of specialization with more elaborate designs. While the dead continued to be buried in underground tombs and caves, the new surface temples gradually acquired a significant role in the religion of the living. Even so, the rock cut tombs were never abandoned. Besides those of Zebbug, others were found in Xemxija heights near St. Paul's Bay in Malta. However, the most spectacular tomb is the oval shaped Holy of Holies, with its semi-circular vaulted roof, carved into the rock in the Hypogeum at Hal Saflieni in Malta. The whole complex is part mausoleum and part temple of worship.

GGANTIJA AND HAGAR QIM

Once introduced, the development of the surtace temples was quick and furious. The Ggantija phase represents the greatest era of the neolithic constructions. Originally the basic plan was of three chambers in a trefoil pattern grouped around a central rectangular court. Soon, this was augmented with massive facades made of huge megalithic slabs, leading to a semi-circular forecourt. The trefoil plan, but on a much enlarged scale, is present in the complex of temples known as the Ggantija on Gozo. Here, lateral chambers are introduced, in addition to altars, niches and tabernacles. The interior decoration becomes more elaborate too. The walls of Ggantija, in some cases twenty feet high and weighing several tons are most impressive. These massive stones were probably transported on stone balls. Some fairly sophisticated hoisting techniques, involving considerable manpower must also have been used in their erection.

Another large and impressive complex of monuments is to be found on the south coast of Malta overlooking the sea with a fine view of the island of Filfla. This group of buildings known as Hagar Qim is not far from the village of Qrendi. The original temple was subject to many alterations and extensions, but the whole complex presents many novelties in construction. There are more entrances, corridors and window-like openings between the various chambers. It is entirely built of the soft globigerina limestone, probably quarried nearby. It also contains the largest single stone used in any of the Maltese temples, measuring 22 feet. A most unusual pillar 'altar', with carvings bearing a floral motif in the lower portion was found at Hagar Qim. This unique motif does not appear in any of the other megalithic structures in Malta.

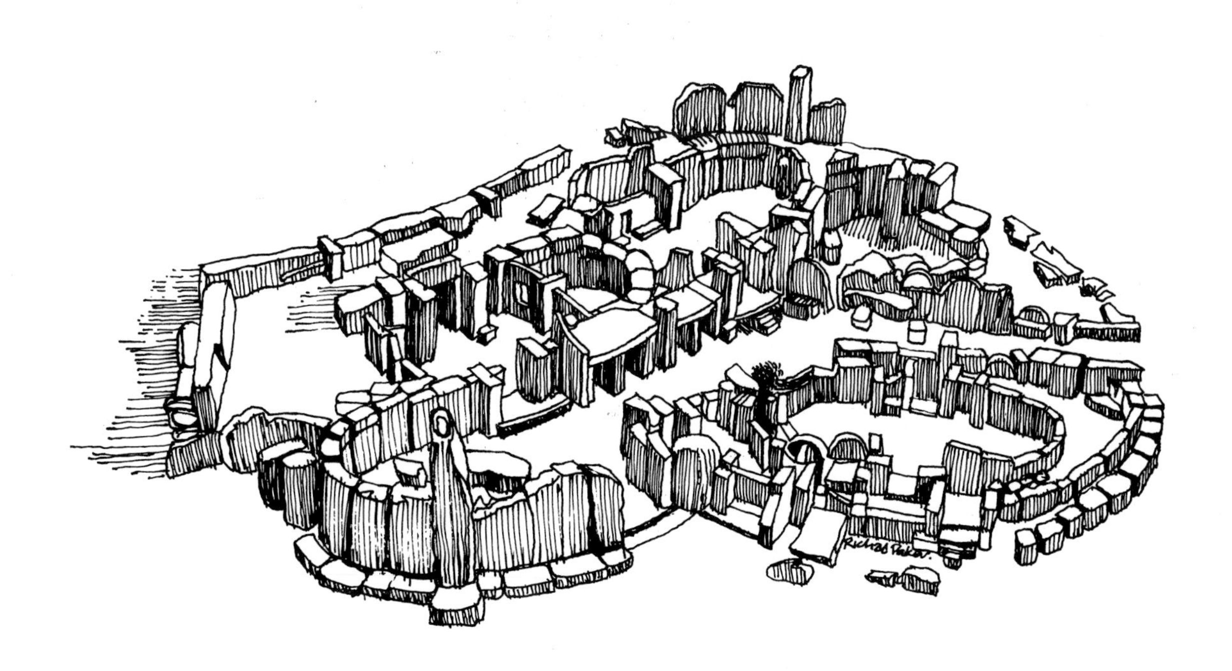

In the light of the lovely moon
all through the sweet hours of the summer nights
there in the hush of the wilderness green for ever
Ggantija dreams:

Dreams that she sees again
out of the few small huts erected near her
two by two issuing, slowly approaching,
the faces she loves.

They enter; suddenly the temple
is filled with a throng of women, children and men,
wan faces; on each face can be read a look
of fear and sadness.

A voice in heard echoing :
'For to dispel the famine that would destroy us
the gods desire forthwith, that they may forgive us,
a maiden to die for us.'

A fearful clamour goes up,
weeping and sobbing, with the murmur of many voices;
The dread of death possesses the hapless multitude
hungry these long months.

But amid all the shouting,
while every voice bewails the grievous tragedy
a maiden, golden tressed, blue of eye, pale
steps out of the throng

And cries in a loud voice:
'Since my arms are of no avail against this
I will give to the gods my life and youth,
I will die for Gozo.'

Then the altar smokes
and there commences the fading anthem of death,
the drums softly rumble out the last lament,
the earth is blood-drenched.

The morrow Gozo was verdant,
our fields were overflowing with fatness and corn,
every one ate and drank, many made revel
every heart rejoiced.

It is of this that, under the moon
all through the sweet hours of the summer nights
there in the hush of the wilderness green for ever
Ggantija dreams.

Gorg Pisani: Ggantija [1]

1. Arberry: Op. Cit. pages 246 - 248

THE HAL SAFLIENI HYPOGEUM

The underground 'temple' near Paola, known as the Hal Saflieni Hypogeum, is one of the fascinating and well preserved prehistoric sites in Malta.

This monument is not strictly a megalithic structure. Although 'built' entirely underground, it has to be considered with the surface temples, the structure of which it reflects and imitates. imitates.

The Hypogeum is carved out of living rock into the top of a natural hill of globigerina limestone. In total, it represents the quintessence of the megalithic art of the surface temples. The only difference, apart from being sub-terranean, is that it served as part temple of worship and part burial place.

This monument is basically a catecomb with a series of rock cut chambers on three levels, interconnected by stairways, corridors and halls. The main entrance is through a built trilithon leading directly to the first and highest level. Surprisingly, this is the oldest and the one largely used for burials. The middle level is the most impressive. Here the chambers are larger and well finished. The walls are elaborately carved to represent the megalithic structures of the later surface temples. There are the same megalithic facades, together with corbelled roofs and halls leading to smaller chambers, with imitation trilithon entrances. The «Holy of Holies» is a small kidney shaped chamber so called because of its dual function as both a burial place and a shrine for rituals and worship. Traces of red ochre; the ritual colour of death, can be seen on some of the facades. Two sleeping lady figurines were found in the middle chambers. The rectangular oracle chamber is also on this level. It is actually made up of three oval adjacent chambers. The smallest one is more like a niche. If spoken into by a deep voice of a man, the echo will reverberate around the chamber. The high pitched voice of a woman will have no such effect. The lowest level is at a depth of some 36 feet. Here the chambers are high and some are extremely narrow. It is said they were also used for burials and possibly for storage purposes. At one end there is a water cistern or 'well'.

The exact function of the hypogeum still baffles archeologists. The extravagance, the care and attention to detail, the quality and finish of some of the interior chambers, especially those of the «Holy of Holies», suggest that the Hypogeum was perhaps the most sacred site of its time. The ways of a society so dominated by the goddess of life and death with its associated cults of ancestor worship and burial rituals, are hard to understand. The remains of some 7000 persons were found in the Hypogeum; this too suggests that it was an important sanctuary serving the dead and the great deity of death - the great Earth Mother.

MNAJDRA AND TARXIEN

The final phase in the development of the megalithic monuments of Malta is best expressed in the temples at Mnajdra and Tarxien.

Those of Mnajdra are not far from their precursors at Hagar Qim and like them, they stand on high ground facing the sea. Here, the outer casing of the walls consist of blocks of coralline limestone. The interior is made up exclusively of the softer globigerina. Though intricate and elaborate in design, the Mnajdra temples perhaps lack the mastery, appeal and prominence of those at nearby Hagar Qim. Because they are beautifully proportioned, they tend to integrate better with their environment. The main portal is ten feet high. It leads to oracle chambers situated behind four apses. Most impressive are the stone columns forming the frameworks of the openings and of the monolithic doors.

Of the three temples at Mnajdra, the oldest, dating back to the Ggantija phase, is called the Small Temple. The lower and middle temples belong to the Tarxien phase. The middle temple is most simple in design and structure and it marks a relative 'decline' from the magnificence of the earlier megalitic structures.

The complex of four temples at Tarxien are possibly the best preserved. The Early Temple dates to the Ggantija phase. The East, Central and South Temples belong to the Tarxien phase. The detail of the temples is at its most elaborate in the carvings to be found nearly everywhere. The basic plan is that of a central aisle with two pairs of apses. The interior is beauifully carved with spiral shapes and animal friezes depicting sheep, goats and pigs. Of tremendous importance is the colossal statue of the Maltese Fertility Deity found in the South Temple. Although only the base remains, standing about three feet in height, there can be no doubt that the original must have stood at no less than eight feet. The statue is quite unique. It is presumed to be the only piece of monumental sculpture belonging to this early period. It was not until several years later that anything like it was attempted in the Mediterranean.

NEOLITHIC MALTA AND HER DECLINE

The society which erected the great megalithic temples left no written records. Nor did it leave any cities. As testimony to its singular achievements, it left only its giant monuments. The fact that there are so many of them in the relatively small area of the Maltese Islands would at first suggest that this society was not only numerous but also capable of harnessing the existing labour resources to achieve its major objectives. This would have involved a high degree of organisation, some specialisation in various skills and a technology of sorts. Yet, we are still most uncertain whether indeed Malta did have a large population. We know next to nothing of its hierarchical system, if that existed. As for its technology, we can safely say that the use of metals was not introduced into the Islands at that time.

In his book 'Before Civilization', Colin Renfrew puts forward the proposition that, according to radio carbon chronology, the temples in Malta are the earliest free standing monuments of stone in the world, [1] even preceeding the earliest pyramid at Sakkara in Egypt. Indeed, it is possible that the temples at Ggantija, Hagar Qim and Tarxien had already been abandoned, pillaged and possibly desecrated by the time the Avebury circles in England, the temple of Mentuhotepb in Egypt or the Ziggurat of Ur, were in the course of construction.

How, and indeed why, so many megalithic temples came to be built in Malta remains mysterious. Indeed, Renfrew compares them to the equally mysterious funerary walled platforms or 'Ahu' on Easter Island in the far Pacific Ocean. He has suggested that both Islands may have had some basic communal social organisation in the form of a hierarchical chiefdom society. He drew up five hypothetical chiefdom territories for Malta and one for Gozo. Each territory, except the one for Gozo, contains roughly two temples in close proximity surrounded by arable land. The territories were largely determined by the sites of the temples. Renfrew also arrives at an estimated population of about 11,000 [2]. It is therefore possible that through the chiefdom scheme of organisation, the labour resources of about 2000 persons in each territory could have been marshalled and directed to erect the giant sacred monuments, so important to their religious beliefs.

The other greater mystery lies in the rather abrupt end of the era of the temple builders at about 2000 BC. The civilization which had enjoyed an uninterupted era of peace for a thousand years or so, suddenly vanished. The Maltese Islands now appear to have been left uninhabited for a few hundred years.

Many theories have been advanced for this singular and curious event. Invasion and complete extermination of the population cannot be entirely ruled out. However, there must have been a more natural cause for this decline and apparent death. It is possible that a combination of factors relating to overpopulation and over-cultivation of limited fertile areas led to malnutrition, death and a rising infant mortality rate. This economic depression may have been gradual over many years. It was certainly not assisted by long periods of drought which started to occur from about 2400 BC. Large scale imports of food were out of the question. Even if such imports were possible, it is doubtful if the Islanders were in a position to pay for them. Basically, Malta was poor and backward. The use of metals was unknown. Apart from its obvious impressive achievements in the building of temples, this society was passive in its nature and perspective. It did not distinguish itself in any other known productive sector. It seemed more obsessed with its religious role. In an economic sense, it was clearly doomed. All the succeeding civilizations in the Maltese Islands have learnt to rely on considerable imports of food and raw materials. Ways had to be devised to pay for them. Neolithic Malta, with its swelling population could not forever remain self-supporting. The seeds of decay were therefore gradually sown from within.

The fact remains that neolithic man, the temple builder of Malta, disappeared from the Maltese scene forever. His great achievements, however, were never to be forgotten. In Malta and Gozo, they still stand as mysteriously passive and as powerful as ever.

1. Colin *Renfrew*: in Before Civilization (Jonathan Cape, London 1973 and Pelican Books 1976). See p. 161 (Pelican Ed.)
2. Renfrew: Ibid. See p. 169

THE BRONZE AGE

TARXIEN CEMETERY PEOPLE

The era of the bronze age civilisation in Malta starts with the new arrivals in about 2000 BC. They brought with them an alien religion and a less refined culture which stemmed from different traditions. This new culture was totally unrelated to that of the temple builders. There is a clear gap in continuity. The one did not evolve from the other.

Although technologically more advanced as a people, it is strange that the pottery of the new settlers is noticeably cruder and plainer than that of the earlier inhabitants. It is different in shape and colour. Aesthetically, it is less refined.

For the first time, copper and bronze daggers, axes and obsidian arrowheads made their appearance on the Maltese Islands. This suggests that the new colonisers, if not more aggressive, at least saw the need to defend themselves. Their settlements were also sited on isolated hill tops which served to emphasise this requirement.

Being of a different religion, they showed no reverence for the earlier giant temples or the cults associated with them. Indeed, they turned most of them into cemeteries where they cremated their dead and buried the ashes in jars or under 'dolmens'. The dolmen was a megalithic structure, consisting of a large capstone, supported by two upright boulders, under which was dug a large chamber to receive the jars containing the ashes of the dead and other burial offerings. Because the temples at Tarxien were used extensively of this purpose by these new inhabitants, they are now commonly referred to as the Tarxien Cemetery people. There are about 20 dolmens scattered over Malta and Gozo. The capstone of the smallest dolmen called 'id-dura tal-mara' in Gozo is 3 feet long. The largest one belonging to the dolmen at 'Misrah Sinjura' in Malta is 12 feet square. Most of the dolmens stand at a height of not more than 3 feet. The Maltese dolmens have a marked similarity to those found in Italy at Otranto. This suggests that these people came from the area surrounding the extreme tip of the heel of Italy. The Tarxien Cemetery People settled in Malta for a few hundred years.

BORG IN-NADUR

BAHRIJA

THE CART TRACKS

It was not until about 1400 BC. that the islands were again invaded. This time, the new arrivals had to fight for possession. They probably came from Sicily and they first established a beach head at Borg-in-Nadur, which dominates the Bay of Marsaxlokk. For this reason, they are known as the Borg-in-Nadur people. They overwhelmed the Cemetery people but it would also appear that the inhabitants integrated within the new culture. There are about eight settlements belonging to this period, including some in Gozo at in-Nuffara, on the slopes of Xaghra not far from Ggantija. Their settlements are always well fortified and, again, on high ground. Although expressing a similar culture, there are sufficient differences in the settlements to suggest that they may have belonged to a succession of colonising expeditions from Sicily. It is possible that these communities also had their political differences which resulted in occasional warring skirmishes amongst each other. Borg-in-Nadur was clearly the most important settlement. Its massive fortified walls also made it the most impregnable.

The Bahrija People were another group of settlers who arrived from mainland Italy in about 1100 BC. They established themselves firmly on the Islandsand especially on the Bahrija ridge to the west of Malta. It would appear that they co-existed well with the Borg-in-Nadur people. They were clearly industrious. There is evidence of a burgeoning textile industry in Malta during this period. The Bahrija people also produced a much finer pottery. Dark and inlaid in white, it is typical of the late Bronze and early Iron ages then prevalent in Sicily and the Italian mainland. The Bahrija phase in Malta continued till about the 8th or 7th century BC, when the Maltese Islands suddenly became of interest to the outside world for the first time. Their strategic location and excellent natural harbours were discovered. With this Malta's prehistoric period closes.

The presence in Malta and Gozo of the so-called 'cart tracks' must now be mentioned. They have been the subject of much speculation and lively controversy. To a certain extent, the tracks remain a mystery.

Basically, they look like cart tracks, consisting of parallel grooves cut into the bare rock. They are V shaped in section. They sometimes cover long distances following various routes along hilly ridges, between valleys and sometimes leading into shallow waters along the coasts. Some of the tracks end abruptly at cliff edges. At some points, they intersect each other. Near the wooded area of Buskett, many of them converge and cross each other. This intersection is commonly referred to as 'Clapham Junction'.

It is now generally agreed that the tracks date back to the Bronze Age period. Most of the settlements and villages of this period have tracks leading to them. It is likely that the tracks were first chipped into the stone on determined routes. Into the V sectioned tracks were fitted 'slide cars' with two parallel shafts shod in stone resting in the tracks. Above the shafts, a platform to carry the load was attached. The friction of the shafts in motion over the stone further deepened the tracks. It is now also agreed that the slide cars were used to carry seaweed from the shore to 'fertilise' the arable land at several destinations. Equally, soil was transported to the more barren areas around the favoured hill top settlements of the Bronze Age people.

Head of a limestone statue menhir from Zebbug-National Museum, Valletta.

Skorba pottery, National Museum, Valletta.

Neolithic temple of Hagar Qim.

The 'Sleeping Lady' from the Hypogeum at Hal Saflieni.

Hagar Qim. ▲

Hal Saflieni Hypogeum - 'The Holy of Holies'. ▼

Hal Saflieni Hypogeum: A small chamber in the lowermost level of the underground temple. Serving as part temple and part mausoleum, the Hypogeum was carved into the living rock. Its lowest level is at a depth of 36 feet.

Limestone headless figure from Hagar Qim.

Bahrija phase pottery, C 1100 B.C., carinated with a strap handle.

Pottery of the early Ggantija phase found at Skorba in Malta.

Page 114: Punic Vase c. 5th Century B.C. A Phoenician/Carthaginian taken into captivity by a Greek. Found in Rabat, Malta.

Hagar Qim, Malta.

Fragment of leather covered with gold leaf embossed with a Phoenician design of two gryphons flanking a multiple palmette with winged disc above. 7th Century BC, Malta.

Terracotta Sarcophagus found near Rabat, Malta. 5th Century BC. National Museum, Malta.

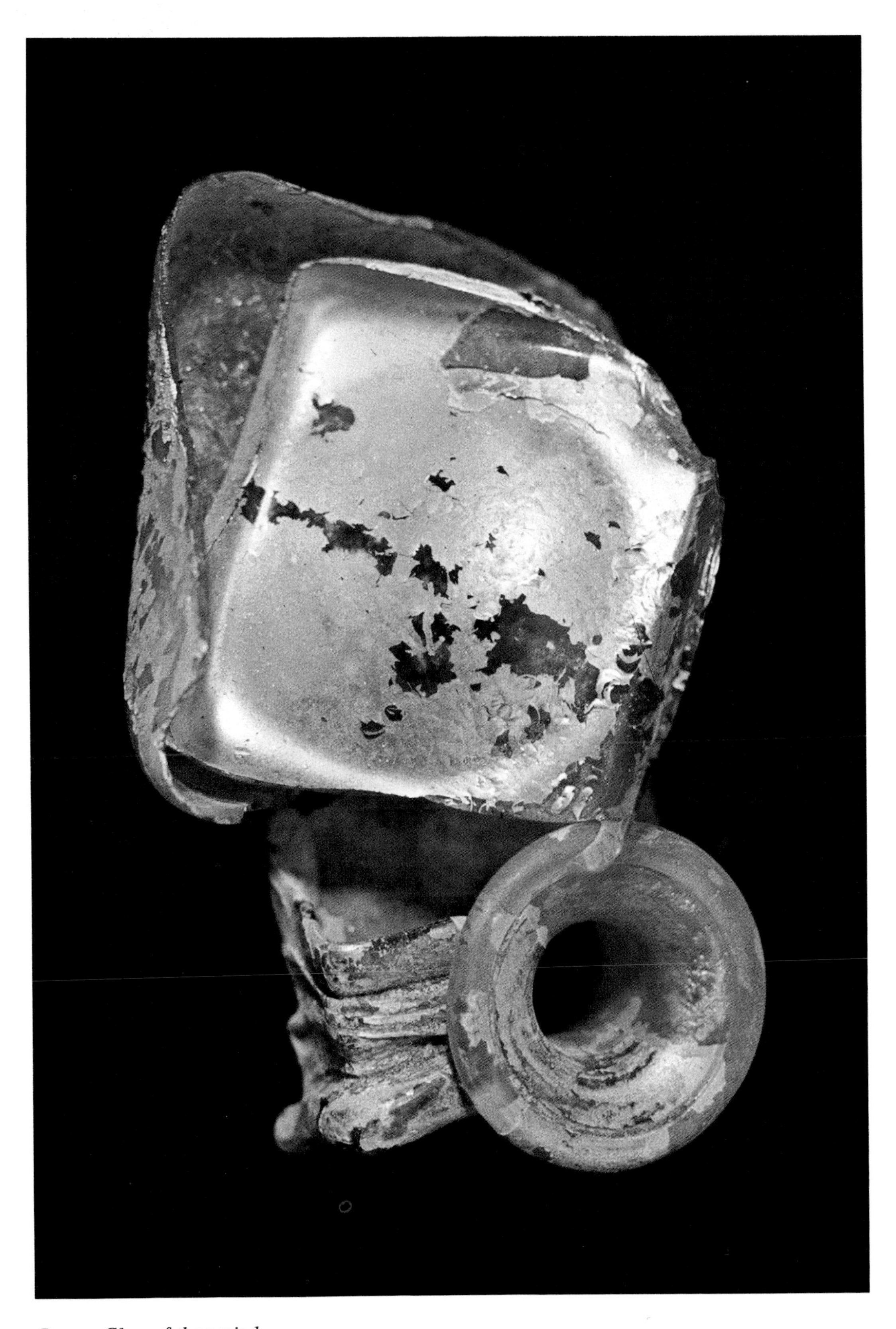

Roman Glass of the period.

Roman oil lamp at Roman Villa and Museum, Rabat, Malta.

Child's bust - *Roman Villa and Museum, Malta.*

Roman mosaics at Roman Villa and Museum, Rabat, Malta.

Page 126: Siculo-Norman period building at Mdina. ➔

THE ENTRANCE OF MALTA INTO HISTORY

PHOENICIANS AND CARTHAGINIANS

The long isolation of the Maltese Islands comes to an end with the appearance of the Phoenicians in the Western Mediterranean. By about 1000 BC, these remarkable people were leaving their imprint along the length and breath of the whole Mediterranean basin. From their homelands in the Levant on the shores of Syria, they took to the sea in their purple sailed ships and ventured deep into the Mediterranean. Along the trade routes to the west which they first opened up, they secured staging posts and planted colonies. Their settlements spread far and wide along the North African coasts and on the islands of Cyprus, Rhodes, Sicily, Malta, Sardinia and Corsica. In time, they reached Spain and penetrated the Atlantic. They then navigated the Bay of Biscay and 'discovered' the tin islands of Scilly and Cornwall.

Sooner or later, the Phoenicians were bound to 'discover' the Maltese Islands. They lay on the direct East-West trade route to Spain. Malta was conveniently located to act as a halfway staging post for rest and replenishment. She was reasonably close to the important Phoenician settlements in North Africa. She was even closer to the colonies in Sicily. Better still, Malta was blessed with the safest, natural harbours of the region. No wonder then, that the Phoenicians called her 'malet' meaning refuge. They used her safe enchorages from the earliest times. Clear evidence of a Phoenician presence in Malta commences with the Eighth Century BC. From this moment on, the punic colony which the Phoenicians founded becomes firmly established. With it, Malta at last enters into the realms of history proper.

The Phoenicians were probably first attracted to the shelter of the great southern harbour of Marsaxlokk in Malta and that of Mgarr on Gozo. The Phoenician word for harbour is 'Qala'. The innermost area of the port at Mgarr is, to this day, known simply as "Il-Qala'. In Malta, a creek within the main harbour of Marsaxlokk is called Qala Frana. It is likely that the word Frana is derived from the word «far» meaning departure in Phoenician. Indeed there is a village called Hal Far close by. It is possible that Qala Frana may have been the port of departure of the Phoenician navies.[1] The Phoenicians did not confine themselves to the coasts. Evidence of their tombs in Rabat and Nigret suggest that they may have founded the city of Melita, now Rabat-Mdina.

It is clear that the bronze age inhabitants of Malta did not resist the intrusion of Phoenician control over the Islands. They probably immediately recognised the cultural superiority of the new colonists. Co-operation with the Phoenicians brought fresh benefits to the islanders. Thanks to the trading 'genius' of the Phoenicians, the Maltese textile industry received a welcome boost as new export markets were opened up to it. Gradually, the culture of the Islanders was assimilated with that of their new masters. There was an increase in population as the Islands entered into an era of relative prosperity.

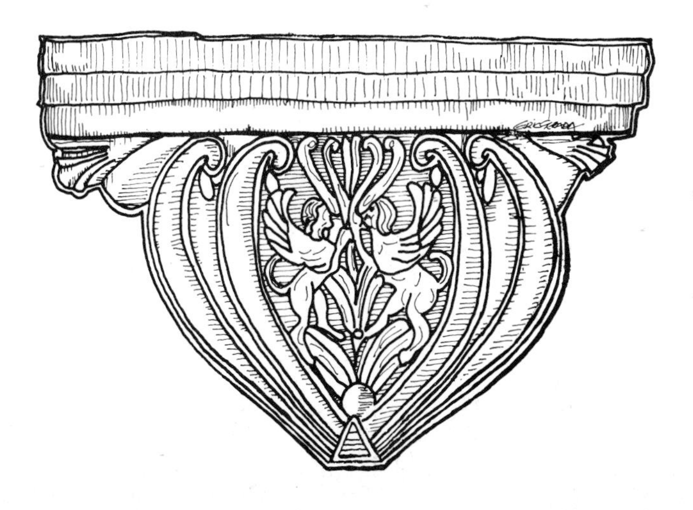

The Phoenicians had many Gods. Every town in Phoenicia honoured its Ba'al or Lord and Master. The Ba'al of Tyre was a solar God in origin, but he was also regarded as a marine deity. He was known under the title of Melkart, God of the city. At Tas-Silg near Marsaxlokk, the Phoenicians built a temple in his honour. It was here that two marble cippi (or Candelabra) were discovered in 1697, they bore inscriptions in both Phoenician and Greek, recording vows to the Lord of Tyre: Melkart.

It is possible that the Greeks too used this temple. They identified Melkart with their own deity Heracles; this was not uncommon since much of Greek mythology is derived from the Phoenician.

The inscriptions on the Cippi enabled the Abbé Bartholemy, an Eighteenth Century orientalist, to decipher the Phoenician alphabet. By the Eight Century BC, this alphabet was adapted by the Greeks who in turn passed it on to the Western world of today. Also from excavations at Tas-Silg, there is now evidence that the Phoenicians built a sanctuary to their Goddess Astarte - the female deity of fertility, beauty and love.

To Melkart, Our Lord of the City of Tyre,
this vow is dedicated by
Abdosir and his brother Osirxamar
both sons of Osirxamar son of Abdosir
hear them and bless them.

Inscription in Phoenician and in Greek
on a marble cippus at Tas-Silg, Malta

Circa 200 BC

Many items of Phoenician pottery were found in the rock cut graves and tombs belonging to this period. They are not dissimilar to those of the Bronze Age people. This suggests that the transition of both cultures was smooth and progressive. The Phoenicians were known to have had elaborate funeral rituals involving dances of grief, with much breast beating and wailing by their women. The burial chambers were adorned with many offerings including plants. Indeed, trees were most important to the Phoenicians. The white poplar was sacred to them and they adorned the temple to Melkart with branches of it. The Phoenician word for tree is 'Luq'. It is possible that the village of Hal Luqa, in the immediate vicinity of the present modern airport, derives its name from trees which may have flourished in the area. Apart from the towns of Rabat in Malta and in Gozo, the other villages which are reputed to be of Phoenician origin are Mellieha, Tarxien, Qrendi and Qormi in Malta, and Ta Qerciem in Gozo. [2]

Punic culture put down deep roots in the Maltese Islands and we can only surmise the impact of the Phoenicians from the development and eventual decline of their power in the western Mediterranean. The influence of the Phoenicians over the whole region and the contribution they made towards its development is underated and, sometimes, forgotten in modern times. They were, after all, the first international traders. To their trading outposts they brought the benefit of an advanced civilisation. Their maritime success was largely due to their supremacy as navigators. They were the first master mariners. As Plutarch commented «They excelled in the arts and in writing and literature as well as in naval warfare and administering the empire».

Although they competed with the Greeks, especially in the eastern Mediterranean, their roles and ambitions differed. The Phoenicians were primarily interested in trade. They sought no territorial dominion. Their colonies were merely outposts along their trade routes. Unlike the Greeks, they were not warlike and they avoided conflict whenever possible. Indeed, they traded well with the Greeks and some of their colonies became partly hellenised. In Malta, the Greeks co-existed happily with the Phoenicians; although they never left any real imprint on the strong punic culture of the islands.

The Phoenicians called themselves 'Chanani' or Canaanites. In the old testament, they are referred to as Sidonians after one of their great cities: Sidon. Their name was probably given to them by the Greeks because of the excellence of the purple dyes produced in the city of Tyre. Indeed, the Phoenicians were commonly known as the 'Purple Men'. It is from the Greek 'Phoinos' meaning blood red, that the name of Phoenicia was derived.

The early Phoenicians settled along the coasts of Syria. Here they founded the great cities of Tyre, Sidon, Arvad and Tripoli. Their influence and trade was always restricted to coastal areas but they were uninterested in conquest or conflict. These people were sailors and not soldiers ; traders not tacticians. They traded in tin, iron, lead and silver; in ebony and ivory; Tyrian purple dyes, coral, honey and spices. They manufactured glass and high quality textiles. [3]

The Phoenicians were the greatest shipbuilders of their time. The areas around Syria provided them with an ample supply of wood ; pines, firs, cypress and the famous cedar from the mountains of Lebanon. Their navigational skills were legendary. From their neighbours in Persia and Egypt they learnt the science of astromomy and combined it to their passion for the sea. Their ships provided security and the sea was their escape. Into the unknown they sailed, always in search of trade. It is said that by the Sixth Century, they had circumnavigated Africa and discovered the Azores. For a long time they were undisputed masters of the Mediterranean.

By the Eighth Century BC, the Greeks had begun to challenge Phoenician supremacy in the eastern Mediterranean. As this pressure intensified, the Phoenicians turned more and more to the west and to their important trade routes to Spain and the tin mines of Britain. In 814 BC, the Phoenicians founded the city of Carthage on the peninsula surrounding Cape Bon in North Africa. Strategically, Carthage with its port, gave the Phoenicians the essential access they required to control their important colonies in Spain and on Corsica, Sardinia and Sicily. It was here that the Greeks made serious incursions into Phoenician territory. To avoid conflict, the Phoenicians withdrew from eastern Sicily but they maintained their bases in the west, at Motya and Lilybeum. The struggle for the total possession of Sicily was to last for five hundred years.

With the Phoenicians engaged in the Western Mediterranean and in North Africa, and with the Greeks consolidating their position in the Aegean, a new and remarkable people now appeared in the central Mediterranean. The Etruscans were a semi-oriental people who had settled along the coast of north west Italy. Like the Phoenicians they were great seamen and able traders. They were also master craftsmen. Soon they were a strong maritime and trading power in the region.

In their advance westwards, the Greeks were not content with their gains in Sicily. Soon they established colonies in Sardinia and Corsica and even reached Marseilles. To the Phoenicians, the Greek presence there was intolerable. It posed a serious threat not only to their colonies in Spain, but also to the land routes, via France, to the tin mines of Britain. In 600 BC, they finally resorted to arms in an attempt to drive the Greeks out. They failed lamentably. In the meantime, the Etruscans were also finding the Greek presence in the western Mediterranean troublesome to their trade and maritime activites in the area. In 535 BC, they formed an alliance with the Phoenicians. Their combined fleets fell upon the Greeks off Corsica forcing them to withdraw from that island and from Sardinia.

Soon the Etruscans were engaged with another Greek threat - this time on mainland Italy. At Cumae, near Naples, they suffered a serious reverse at the hands of the Greeks. From this moment on, the Etruscans started to decline. The Greeks compelled them to withdraw from all of southern Italy. By 510 BC, the Etruscan Kings in Rome had been overthrown and Rome was declared an independent republic. Soon it would begin to expand its influence and assume power over all of Etruria and its dominions.

The Phoenicians also start to fade from history in about 600 BC with their defeat in Tyre and Sidon by the Assyrians and Babylonians. It is their successors, the Carthaginians who now assume the mantle of leadership of Phoenician interests in the central and western Mediterranean. At about this time, the Carthaginians took possesion of Malta. Immediately they recognised its strateigic position as a defensive outpost. From here, they could harass the Greeks in Sicily and deter them from advancing south and west to the seat of their power - the great city of Carthage.

It is believed that under the Carthaginians, the Maltese were allowed a restricted form of self government and the islands minted their own currency. The Islands also had a Senate and an assembly of the people. Although never a Greek colony, it is clear that there existed a strong Greek influence in Malta. Many Greek items of pottery, inscriptions and coins belonging to this period have been found.

For a time, the Maltese Islands were largely unaffected by all that was happening around them. Indeed, the only disturbance was more likely to come from armed pirates, then roaming around most of the defenceless islands in the Mediterranean. Their raiding parties would suddenly swoop in and carry the unfortunate who were captured to the slave markets of the great cities.

Relations between the Greeks and the Carthaginians continued to deteriorate. By the Fifth Century BC, the backlash of anti-Greek reaction is felt even in Malta. A gradual decline of Greek influence is noticeable in the poorer quality of craftsmanship and finish of items of artistic merit produced in Malta during this period. The temple at Tas- Silg shows little or no progress for the next two centuries. It is not till the Third Century BC, that Hellenistic influences start to reappear at Tas-Silg. Here, the sanctuaries are now dedicated to Hera, the Greek goddess of marriage and fertility and wife of Zeus. Tanit, the Carthaginian goddess of the sky and fertility, replaces Astarte.

During the Fourth and Third centuries BC, the Carthaginians and the Sicilian Greeks remained locked in continual battle. As always, Sicily remained the main arena of conflict. Indeed, Sicily was finally to be the main cause for the defeat and decline of Athens at the hands of the Spartans in the Peloponnese Wars. This led to the emergence of the Macedonians and of Alexander the Great, who directed Greek interests towards Egypt and the near East. In the meantime, with the decline of the Etruscans, the shadow of Rome over Italy was moving steadily southwards.

It was indeed the Romans who were to fill the vacuum of power in the central Mediterranean - but first they had to contend with the power of the great commercial empire of Carthage, in the Three Punic Wars from 264 BC to 146 BC. For over a hundred years, the Romans and the Carthaginians engaged each other in a death struggle which was to culminate in the final two-year siege of Carthage. In 146 BC, this great Punic city, the «pearl» of semitic culture and power which dominated the Mediterranean for seven centuries, fell to the Romans who razed it to the ground, completely and thoroughly.

Rome became supreme in the central and western Mediterranean. Soon, after yet another bloody struggle with the Greeks, the Romans established themselves as the complete masters of a unified Mediterranean. The era of «Pax Romana» had arrived.

1. Dunstan G. *Bellanti*: in Why Malta Why Ghawdex (Orphans Press, Gozo 1934). In this publication, Bellanti suggests the derivation and origin of the names of the towns and villages of the Maltese Islands.
2. Bellanti: Ibid.
3. Ernle *Bradford*: in Mediterranean Portrait of a Sea (Hodder and Stoughton, London 1971) p. 67.

ROMANS AND BYZANTINES

To the South of Sicily are three islands each of them with towns and harbours offering shelter to all ships cast thither by storms. The first is Melita, 800 stadia from Syracuse having many convenient ports. The inhabitants abound in opulence, for they have artifices for every kind of work; but they excel most in their manufacture of linen which is beyond anything of the kind, both in firmness of its texture and its softness. Their houses are very beautiful and magnificently ornamented with pediments projecting forward and the most exquisite stucco work. The inhabitants have become very wealthy and increased in reputation and splendour. [1]

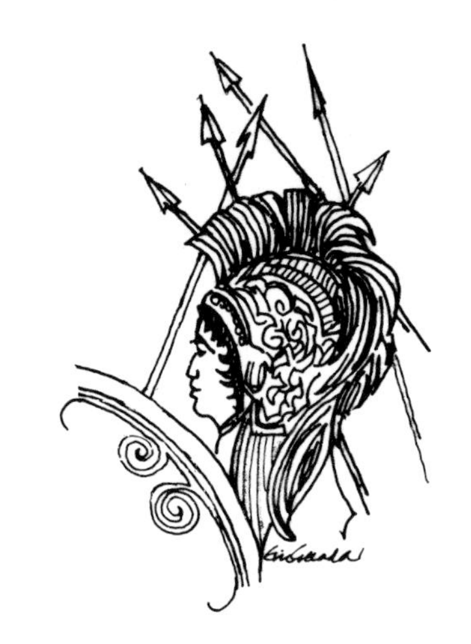

Diodorus Siculus : Circa 40 BC.

The great sea battles between the rival navies of Rome and Carthage commenced with the First Punic War in 264 BC. In support of the now urgent requirements of their navy, the Carthaginians had established a strong naval base in Malta. Here they enjoyed the logistic facilities to be found in the sheltered natural harbours. The other great advantage they enjoyed over the Romans lay in their superior navy which was largely composed of quinquiremes: the five-banked battleships of over 300 oars which were superior to anything afloat in the Mediterranean during the First Punic War. No doubt, the Romans copied them later. [2]

The Romans had long realised the strategic value of Malta. There is evidence that they attacked the Islands in 257 BC, and possibly captured them. They did not enjoy continual possession for long and it is likely that Malta alternated between the hands of the rival powers. In 241 BC, a relative lull in the war came about. A treaty was concluded whereby Sicily was ceded to the Romans, with Malta being retained by the Carthaginians. This uneasy peace did not last for very long. The Second Punic War broke out in 218 BC. It was clear to the Romans that, in order to minimise the harassment by Carthaginian shipping, they had to win control of the central Mediterranean. Besides Sicily, Malta had to be taken, and very quickly. From Lilybeum (Marsala) in Sicily, itself an ex-Phoenician and Carthaginian stronghold for many years, Titus Sempronius crossed over to Malta. Here, he secured the surrender of the Carthaginian commander Hamilcar, son of Gisco and his 2000-strong garrision. Whether the Maltese assisted in the invasion or not is still in doubt. At first, the Romans considered Malta as a conquered territory governed by Rome. Indeed, it is likely that the most senior person in the Islands was a 'Procurator'. He was responsible for both military and civil affairs. Because he was chosen from among the local elite directly by Caesar himself, the Islands were regarded as an Imperial province ; unlike Sicily, which was known as a Senatorial province. The Islands certainly minted their own currency at an early stage. It was not until the beginning of the Second Century that Malta and Gozo were raised to the level of a 'Municipium' with autonomy of local government vested with legislative, executive and, to a lesser extent, judicial functions.[3] Malta was not a bed of roses during the early phase of the Roman occupation. Like all Roman subjects, at a time when Rome itself had not acquired much of its later sophistication, the Maltese were merely tolerated and tactfully 'admitted' into an alliance with Rome. Malta retained its strong punic traditions and language for many years. It was only after a long gradual process that the Islands became 'Romanised'. Certainly, things improved over the years. As Roman laws were introduced, the Islands settled comfortably into an efficient administration system under the protection of 'Pax Romana'.

In a colonial sense, the Maltese started to emulate their masters and adopt many of their social norms. Their manners became romanised and some Maltese even took Roman names. There is evidence of this in the inscriptions discovered at Mtarfa where a temple dedicated to Prosperine was erected in about 23 BC. The Punic flavour of Maltese customs and language were being reshaped and interlaced within the Roman idiom. Gradually, the culture of the Graeco-Latin civilization took root. Its attendant life-style spread with it. The newly found prosperity of the Maltese was being reflected in the same lavish and sumptious villas of the Romans which Diodorus Siculus was so impressed by. Despite the ups and downs, this wealth must have engendered a measure of goodwill towards Rome and the new Latin civilization. The general progress and development of Malta within the Roman world must also have further raised the hopes, ambitions and confidence of the new Maltese.

In Malta, the Romans developed the port facilities in the Marsa area, in Burmarrad, Salina and at St. Paul's Bay. In Gozo, the harbours at Xlendi and Marsalforn were also in active use. This led to an expansion of trade, as new export markets were opened up for the existing textile industry, now famous throughout the Roman world for the best sail cloth, linen and garments. Equally famous was Malta's honey. The Romans may well have called Malta by the name of 'Melita' from 'Meli' meaning honey.

There were advances in agriculture. Irrigation methods and water supply were improved. There was an increase in the production of wheat and olives. Indeed, most of the Roman villas or country houses, usually well situated on high ridges overlooking the fertile valleys, had their own 'olive pipper'. These rather large 'gadgets' were made of stone. In a grinding action, they separated the pulp from the olive stone and extracted the oil.

Although Malta prospered, it is unlikely that she ever enjoyed any significant status within the Roman scheme of things. Her strategic value was diminished once the Mediterranean was more or less unified. The Islands became mere provincial outposts of Rome, possibly serving even as an exile station. However, it has to be said that Cicero did express a desire to retire in Malta; although he never did. Once here, the Romans did not find Malta unattractive. The villas, the baths, the fortifications, the temples and tombs, the inscriptions and the artistic relics are all evidence of progress, prosperity and of a fairly high cultural level.[4]

In their religious life, the neo-Punic Maltese and the Romans were not at variance. The old sanctuary of Tas-Silg was enlarged. The Maltese now identified their goddess Tanit with the Roman Juno, Ba'al Hamoon with Saturn and Melkart with Hercules. It is said that a temple to Apollo was erected in the city of Melita, now Rabat-Mdina. Temples to Juno existed between Fort St. Angelo and Birgu in Malta, and at Gaulos, now Rabat in Gozo.

A good indication of the smooth relations existing between Malta and Rome can be gleaned from an incident which occured in 70 BC. The Maltese sent their own delegates to Rome in order to denounce the Roman Provincial Governor of Sicily and Malta, Caius Verres, for plundering and desecrating various temples. When brought to justice, Cicero accused him of robbing the temple of Juno of many votive offerings housed there, including some sacred ivory tusks. Verres was further indicted for generally ransacking all that he could lay his hands on during his three-year rule over Sicily and Malta. He was accused of thieving «Four hundred jars of honey, a great quantity of Maltese cloth, fifty cushions for sofas and that number again of candelabra and garments.» Caius Verres was found guilty and he probably went into exile.

The most important single event to occur during the Roman occupation was the shipwrecking of St. Paul in A.D. 58. With St. Luke, he was on a voyage from Caesarea to Rome as a prisoner of a centurian named Julius. Their ship struck rock on one of the islands at the entry to St. Paul's Pay, so named after the great Apostle. This event is well recorded by St. Luke in the Acts of the Apostles. In Malta, St. Paul was well received by the Maltese and by Publius, the leading Maltese citizen. It is said that, he spent three days with Publius in his villa at San Pawl Milqi, in Burmarrad. Thereafter, St. Paul is reputed to have lived in a grotto at Rabat. He stayed in Malta for three months since, because of bad weather, not much shipping took to the sea during the winter months in the Mediteranean.

St. Paul preached the Gospel with his usual zeal. After allegedly curing the ailing father of Publius, it is said that he secured his conversion to Christianity. Later on, Publius was appointed Bishop of Malta. During this period, the first seeds of Christianity were sown in Malta. The Islands thus became one of the first Roman dominions to embrace this young religion.

The emergence in strength of Christianity in Malta was not as sudden as legend would have it. It was certainly not accomplished during the three month visit of St. Paul. Indeed, the Saint must have suffered a few reverses here and there. Legend has it that the villagers of Zejtun adamantly refused conversion. They stamped their feet firmly as St. Paul preached. They even assaulted him. St. Paul's reaction was to lay a curse on the villagers, whereby they would forever be 'flat footed'. For this reason, the people of Zejtun are known as «Ta' saqajhom catta» or «the flat footed». [5]

In Rabat, the Grotto of St. Paul can still be seen along with the largest complex of Christian tombs known as St. Paul's Catacombs. The early Christians used these extensively as burial tombs and as shrines. When Christianity was more widespread, more tolerated and finally institutionalised, churches and chapels were built over the catacombs. The church of St. Catald opposite the Grotto of St. Paul is a good example. Not far, can be seen the catacombs of St. Agatha, noted for their arcosolium tombs dating to the 4th - 6th century. It is said that St. Agatha lived here after fleeing from the Roman persecution of Christians in Sicily.

It would appear that by the Third century A.D., Christianity was the accepted religion among the majority of the population. Possibly, this may have soured relations with the Romans who were themselves being infiltrated from within by various religious sects. In their dominions to the East, Mithraism was increasingly popular among Roman soldiers. At first, the Romans tolerated Christianity in Malta and elsewhere. Their chief enemy was Judaism. The Jews were severely persecuted. Jerusalem was sacked twice by the Romans; first in A.D. 70 and then in A.D. 135 when it was completely annihilated. Roman persecution of Christians came later, when the uncompromising qualities of Christianity were creating many hero martyrs for the Faith and many 'conquests' in the East. By the Third century, the Roman Empire was fully extended and becoming increasingly dependent on the locally recruited armies of its various dominions. It was only a matter of time before the powerful Roman possessions in the east should be converted to Christianity. The death knell of pagan Rome was finally sounded early in the Fourth century when the Emperor Constantine became Christian himself. With renewed vigour, he embarked on the creation of a 'Christian Rome' on the banks of the Bosphorous. Constantinople was born.

Little is known of the repercussions which these events may have produced in Malta. In A.D. 330, when Constantine was busy shaping his new city, Malta was very much a backwater, but she was still enjoying a measure of prosperity which she was not to see again until the golden era of the Knights of St. John, many centuries later.

Despite the long period of Roman rule, comparatively few Roman antiquities can be seen in Malta today. No doubt, there is still much to be excavated. However, the Roman Villa at Rabat houses some interesting relics and some beautiful and well preserved mosaics. Built over the original Roman Villa stands a museum where many items recovered from tombs and other sites can be seen. A good example of the large 'olive pipper' used in Roman times is housed in the museum. The best example of Roman baths or 'thermae' in Malta can be seen at Ghajn Tuffieha - the spring of the apple tree. Here, the mosaic floors are also worthy of mention. Although many country houses are known to have flourished during Roman times, very little is left of them. The most important is that at San Pawl Milqi - St. Paul welcomed - in Burmarrad.

By the end of the Fourth century, it is likely that Malta came under the jurisdiction of the Roman Empire of the East based on Constantinople. At this point, the Maltese were enjoying more freedom to practice their Christianity. We now find inscriptions bearing many Christian symbols, in the shape of the traditional 'fish', the palm of peace and the 'cross and book'. The first chapels were also being erected.

Turbulent times were to follow in the saga of the Mediterranean. The Vandals and the Goths were sweeping down from the North. Soon, Roman Spain fell. It was followed by Mauretania, Numidia and finally Roman Africa. The devastation and plunder for which the Vandals became notorious was the scourge of the times. By the middle of the Fifth century, they were virtual masters of the Mediterranean. In A.D. 455, they took Rome. For another hundred years, the Vandals pillaged and shook every known foundation on which Latin civilization had been built.

There is some doubt as to whether or not the Vandals actually took possession of Malta. Certainly, no trace of them exists. «It seems probable that towards the end of the Fifth century, as Roman domination itself evaporated, Malta came under the distant control of the Ostrogothic rulers who succeeded the Vandals in Italy, and that it then passed into the Byzantine sphere of influence.» [6]

Again, the Byzantine period is shrouded in doubt since so little is known of it in Malta. One can only speculate on the conditions then prevailing in Malta. It is likely that Christianity came into its own during this period. From the sites where Byzantine pottery was discovered, it is possible to conclude that a few churches were built, the most important of which may have been a basilica with a baptistry over the old sanctuary at Tas-Silg. At San Pawl Milqi, there is evidence of Christian occupation. It is said that by the late Sixth century Malta had its own bishop. «There may have been monasteries on Malta and a papal estate on Gozo.» [7] The evidence of strong Byzantine influence on the Church and within the ecclesiastical organisation via Sicily is clear.

Of the early Byzantine connection, all one can be certain of is that Bellisarius, the great general of the Emperor Justinian did mount a rescue operation in the Mediterranean on behalf of the Byzantines. He managed to salvage parts of North Africa and on his way there in A.D. 533, he called on Malta. Whether the Islands needed to be rescued or not is still unclear. In A.D. 535, the Byzantines conquered Sicily.

In time, Byzantine Malta may have become a strong naval base. Because of the growing threat of the Berbers in North Africa, which indeed prompted the exodus of Christians from Tunis, and the increasing appeal and militancy of Islam, Malta was to regain some of her strategic value. It is possible that the system of Roman lookout towers on Malta's coastline was re-activated and extended. Malta's defence was important to the Byzantines who were still largely in control of the Mediterranean sea lanes. By the time that Islam took to the sea in the Seventh century, Malta was probably well fortified and garrisoned. Indeed, she held out to many an Arab incursion in the Ninth Century. When Malta finally capitulated to the Arabs in A.D. 870, she did so after much resistance.

The period of Graeco-Latin rule and influence through the Romans and the Byzantines thus came to an end. Possibly because it was largely peaceful, the Roman period has tended to remain obscure. The scarcity of excavated sites diminishes their achievements in the Islands. Roman influence in Malta's development and particularly in the future outlook of the neo-Punic Maltese is often underestimated. It has to be considered that the Roman era in Malta was one of the longest periods of single foreign power domination in the history of the Islands. It lasted over 600 years. Indeed, through the Byzantine connection which followed it, the influence of the Romans in Malta, through their Latin civilization, may well have lasted for a thousand years.

The Roman period represents the first contact of the Maltese with the new Latin civilization, which fired the emergence of modern Europe. The emotional links of the Maltese with Europeans were first enkindled during this time. The Romans left their stamp in Malta, not only in the order, organisation, legislation and systems for which they were noted; but also to a greater degree they influenced and reshaped Malta's artistic heritage and cultural traits with the introduction of their Graeco-Latin art forms and civilisation.

It is correctly assumed that the long Semitic and Punic domination of the Maltese Islands before Roman times has left certain indelible features which may have blossomed again with the next phase of Semitic Arab rule which commenced in A.D. 870. To this day, these influences prevail, particulary in the Maltese language, place names and in elements of folklore. By the same token, the full impact of the Euro-Roman, possibly over a longer period of a thousand years, cannot be under-valued. With the next phase of European, and now Christian domination, which occurred with the arrival of the Normans in A.D. 1090, it is also possible that the Graeco-Latin elements of the Maltese personality and culture were re-vitalized.

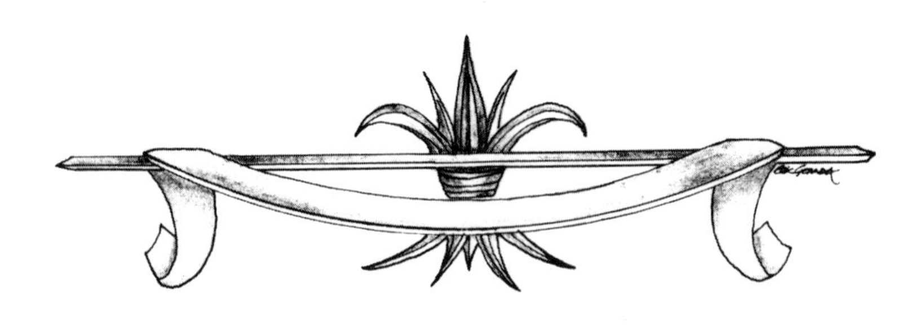

References:

1. Quoted in Ancient Malta, A study of its Antiquities (Colin Smythe, 1977) by *Harrison Lewis* p. 95.
2. Bradford: Op. cit. p. 201.
3. Andrew P. *Vella*: Storja ta' Malta Vol. 1 (Klabb Kotba Maltin 1974) p. 38.
4. Anthony T. *Luttrell*: Approaches to Medieval Malta (The British School at Rome, London 1975) p. 20.
5. Joseph *Cassar-Pullicino*: Studies in Maltese Folklore (Malta University Publication 1976) p. 128.
6. Luttrell: Ibid. p. 21.
7. Luttrell: Ibid. p. 21.

THE ARABS AND THE NORMANS

In the name of God the Compassionate and Merciful.
God look kindly upon the Prophet Muhammad and his family, and grant them eternal well-being.
To God belongs power and everlasting life, and his creatures are destined for death; And you have a fine example in the Prophet of God.

Koran, xxxiii.21

At the close of the Sixth century A.D., old Arabia was in the throes of disorder, anarchy and idolatry. To the north, this conglomeration of ineffective states was menaced by the rival empires of Byzantium and Persia; to the south, by the Abyssinians and the Persians.

The nomad Arabs were originally animists by religion. They worshipped natural phenomena such as trees, rocks and water springs. In time, they had developed a polytheism based on a variety of spirits, but still fired by their animist beliefs. The more settled agricultural communities, in their gradual contact with Byzantines, Abyssinians and Persians were slowly advancing towards monotheism. Indeed, to the South, a belief in the Supreme God 'Al Rahman' - the Merciful, was now already well rooted.

The great monotheistic religions of Judiasm and Christianity, although locked in strife, were clearly on the ascendant in the whole region. Arabia, now a political and spiritual vacuum was ripe for leadership and unification. The Arabs were also ready to discard their polytheism and their petty idols and embrace a new monotheistic faith. For greater impact,this faith had to be so tailor-made and assembled to take into account the strengths and weaknesses of the total Arab personality and also embrace the tenets of the one merciful God.

The leader to accomplish all this, and more, was born in Mecca in A.D. 571. From early childhood, following a meeting with a Christian monk, the great prophet Muhammad was obsessed with the urgent need for one God to be recognised by the Arabs, and for a prophet to emerge and proclaim Him within Arabia. It is said that in A.D. 612, the angel Gabriel charged Muhammad to do just that. Islam, after the Arabic word for 'submission' to the one great God, was born. The new faith swept the idols off the face of Arabia, first in Medina and then in Mecca. With a large army of militant followers, Muhammad besieged and finally unified greater Arabia under the banner of Islam.

The sword is the key of heaven and hell. All who draw it in the cause of the Faith will be rewarded with temporal advantages. . . If they die in battle, they will be transported to Paradise, there to revel in eternal pleasure in the arms of black eyed houris.

Koran, xlvii, 15

Despite Muhammad's sudden death in A.D. 632, the compelling call of Islam fired his united armies to go forth out of Arabia and conquer Palestine, Egypt, Syria and Iraq. Within a hundred years of the Prophet's death, the Muslim empire had spread east of Syria and west into Africa. After taking Spain in A.D. 714, the Arabs advanced through the Pyrenees and into France; only to be stopped at Poitiers by the Frank, Charles Martel in A.D. 732.

The Arab subjugation of North Africa was not accomplished overnight. It took them nearly a hundred years finally to dominate the equally nomadic Berbers. Once this was achieved, the process of Arabization and Islamization was perhaps the most thorough and complete. As the Christian Latins and Greeks 'withdrew' to Spain and Sicily, and some possibly even to Malta, the indigenous Berbers embraced the unifying creed of Islam. With it, Arabic took easy root within the original Punic and Semitic tongue of the Berbers. Kairouan and Fez, the great Arab-Berber cities in Tunisia and Morocco became centres of leadership of the former Roman Africa, Numidia and Mauretania.

From Kairouan, the great Aghlabid dynasty emerged in strength. The horsemen of the desert soon took to the sea and became seasoned sailors and navigators. After invading Sicily they captured Palermo in A.D. 832. From here, they crossed into mainland Italy. Sweeping north, they soon took Ostia and then sacked St. Peter's Shrine in Rome. By the Ninth century, they were in virtual control of the Mediterranean.

It is possible that the Aghlabids started to raid the Maltese Islands as early as A.D. 836. [1] In 869, they besieged Malta and failing to take Mdina, they were again driven out by the Byzantines and the Christian population. A year later, from Sicily, the Aghlabids sent their fleet under the command of their admiral Ahmad bin-Umar known as Habasi. [2] He met strong resistance from the Byzantines and the Maltese. Finally, they were overwhelmed. After a thousand years, Malta again came into contact with a new and vigorous semitic race of people.

The Mediterranean, and Malta with it, benefitted enormously from the invigorating spitit which came from the East. With the collapse of the western Roman empire and the devastation by the Vandals and the Goths, that followed in its wake, the region was ripe for creative change and very much in need of fresh ideas. These the Arabs provided in abundance. With them they brought the accumulated wisdom of centuries-old learning of Syria and Persia, of hellenised Alexandria and ancient Babylon. In its militancy, Islam had assimilated and improved on the knowledge of astronomy, mathematics, science and medicine which it acquired in the course of its great invasions. Into the Mediterranean, they introduced new art forms. Far from threatening the culture of Christian Europe, they revitalized it.

It is a sad (and curious) fact that very little documentation relating to the two centuries of Arab rule in Malta survives today. Indeed, Arab influence in Malta lasted for much longer; since the relatively enlightened Normans who followed them tolerated the extension of their presence on the Islands. A few Arabic inscriptions in Kufic script bearing quotations from the Koran along with some names and places were found; but little else.

It would appear that the Arabs concentrated on the Rabat-Mdina area where many Muslim graves have been found. In one of them, over the site of the original Roman villa, a solid silver ring with the inscription «Rabbi Allah Wahid» meaning «My Lord is One God» was also discovered. [3] The Arabs fortified the city of Mdina. The walled bastions now seperating it from Rabat are ascribed to them. It is also possible that they built a mosque at Tas-Silg. At San Pawl Milqi there is evidence of their occupation.

If the hypothesis that the Islands became predominantly Muslim during the period of Arab rule is correct, it is strange that no further evidence of their mosques exists in Malta.

The Arabs were renowned for their poetry and their literary pursuits. They encouraged literacy and the arts everywhere. Wheras reading and writing were the preserve of the clergy and a few isolated professional scribes in Christian Europe, literacy flourished in Arab Cordoba and in Palermo amongst other places. In Cordoba, they built a great university and the city boasted of many libraries. Although it may be presumptious to expect to find parallel examples of Arab achievements in Malta, it is nonetheless sad that even lesser examples are non-existent in the Islands. Very little is written about Malta by Arab writers. The little information that exists on three Maltese Arabic poets is also very confusing. Despite the enormous impact which Arab art forms had on many parts of the European Mediterranean and elsewhere, and particularly in the field of ornate architecture which can still be seen in Spain and Sicily, its singular absence in Malta is misleading and curious. It seems certain that Sicily was the 'hub' of the Aghlabid, and later Fatimid Arabs' creative activity. The fusion of Arab culture within the Graeco-Latin heritage of Sicily yielded marked advances in many fields. Malta was very much the poor relation and the Islands, although useful, probably remained a backwater in the Arab scheme of things.

The chief legacy of the Arab occupation in the Maltese Islands must forever remain in the Maltese language, which in its roots, is closely akin to Arabic. Here the Arab imprint is indelible. Their penetrating influence is also felt in the names of many villages in Malta and Gozo like Mgarr, Msida, Mqabba, Ghajn Sielem, Ix-Xaghra and Zurrieq. Indeed the names of Filfla and Comino are also of Arabic origin. [4] That the Maltese so completely assimilated Arabic into their mode of speech is clearly due to their ready understanding of, and acquaintance with, the semitic tongues of past Punic times.

The relationship between Muslims and Christians, especially during the early years of Arab domination is a controversial subject. It is generally presumed that, as in Sicily, the Arabs were tolerent of the Christians in Malta. Because of the clear Maltese involvement in assisting the Byzantines to resist the Arab invasions and final take-over, it is possible that the Maltese may have suffered some persecution at first. The Maltese were clearly encouraged to embrace Islam, but the practice of Christianity was never banned. Muslim converts were not liable to the special tax imposed on Christians. No doubt, this may have caused some defections among the Maltese.

It is said that in A.D. 878, the Arabs expelled the local bishop and possibly some of his supporters, and imprisoned them in Palermo. They did this on charges of treason for having collaborated with the Byzantines during the Arab siege of Syracuse. Whether or not this drastic act was directly intended to suppress Christianity, we can never be sure. However, although the bishop was later freed [5], he was never re-instated; nor was a successor ever appointed during the period of Arab occupation. More than likely, the Arabs merely wished to minimize the formation of a rallying point of potential opposition to them, which may well have existed among the Christian population. It is also said that the Arabs looted some items including some carved stonework and marble pillars from the sanctuary of Prosperine at Imtarfa and possibly from the Byzantine church at Tas-Silg. These ended up in the Emir's palace at Sousse, near Tunis [6]. With time however, and early persecution of Christians in Malta evaporated, and it can be said that the level of Muslim tolerence of Christians - and indeed of Jews - was probably high. It certainly does not compare with the totality and severity of Christian persecution of Muslims and Jews in later years.

Under the Arabs, Malta was administered by a Governor or Qa'id in consultation with the leading citizens, who were, no doubt, Muslims. As in most parts of the Mediterranean, agriculture in Malta flourished as a result of the more advanced irrigation techniques which the Arabs introduced. They pioneered the use of the Maltese «Sienja», an animal powered mechanical device for drawing well-water on to land[7]. It is said that the Arabs introduced the cultivation of cotton, which was to become the mainstay of rural Malta for a long time. From India and the east, the Arabs had «discovered» citrus fruits and these too, they introduced to the Mediterranean, including Malta. The agricultural scene was thus much enhanced in the Islands. Indeed, this uncanny Moorish flavour still lingers on in Maltese farms and gardens to this day. It bears testimony to the strong Arabic influence and interest in agriculture.

As in Punic times, Malta benefitted from her key position in the central Mediterranean. Again, she was acting as a staging post between Sicily, the «pearl» of the Aghlabid Arabs and their seat in North Africa. In AD 909 the Sunni Aghlabids in Tunisia were everthrown by Said Ibn Hussein, the founder of the Fatimid dynasty belonging to the Shia sect. On the Tunisian coast near Kairouan, he built a new capital and called it Mahdija. From the Aghlabids, the Fatimids inherited Sicily and Malta, where Muslim rule continued with renewed vigour.

By now, the Arabs were the leading merchant mariners in the central and western Mediterranean. In Malta, a thriving export trade was developed. It is also possible that together with the Arabs, the Maltese were heavily involved in acts of piracy, not unconnected with the lucrative slave trade then florishing everywhere. Indeed in Malta the number of slaves exceeded that of the free population. The Maltese took easily to the sea and they were soon providing some of the ablest corsairs. One curious legacy which may be attributed to Moorish sailing techniques is the sailing rig of the «Gozo boat» which is not dissimilar to the Arab «dhows»[8]. The Arabs are also reputed to have first made use of the magnetic compass in the Mediterranean [9] .

Arab hegemony of the Mediterranean started to decline with the entry of the Normans into its power struggles. Of Scandinavian origin, these «Northmen» had, over the years secured and colonized that part of France known as Normandy. Although basically warlike and always bent on conflict and conquest, the Normans were wise enough to recognize superior values and cultures of others in the lands they invaded. They were always prepared to improve themselves and to adopt the superior tenets of the civilizations they overcame. With great ease, they assimilated themselves within these new environments. In Normandy, they adopted the French language to the exclusion of their own and introduced it wherever they went. They also embraced Christianity and the French way of life.

Not content with taking England in 1066, the ambitious Normans were at the same time penetrating the Mediterranean and challenging Muslim supremacy in the region. With the encouragement of Pope Nicholas II, the two sons of Tancrede de Hautville, Roger and Robert Guiscard were soon in control of Southern Italy. Their campaign in Sicily met fiercer resistance and it lasted for nearly thirty years. First Messina, then Catania and finally Syracuse fell to the Normans. By 1090, they were in total possession of Sicily. In that year, they also attacked Malta.

In Sicily, the Normans found a fascinating mixture of Greeks, Latins and Arabs. The prevailing influence of Islam made a lasting impression on them. Instead of suppressing it, Count Roger was enlightened enough to profit by it. He patronized the new culture, if not its religion. He retained the Muslims in government and generally allowed the cosmopolitan character of the population to blossom. There was freedom of thought, religion and enterprise. The Mediterranean was taking a different shape. Although Islam was on the wane, Spain too was still benefitting from the advances made by the enlightened Muslim Omayyad rule. In the central Mediterranean, the Normans, with the clear support of the Popes, were acting as harbingers of militant Christianity. To the east, the Seljuk Turks were leading the way for a new Islamic power to arise; that of the Ottomans.

It would appear that the Norman invasion of Malta and Gozo in 1090 was merely intended to strengthen and consolidate their gains in Sicily. To the Normans, Malta, with the best natural harbours in the central Mediterranean was strategically an extension of their hold on Sicily. A possible regrouping of Arabic forces in an island in such close proximity to Sicily could never be tolerated. The raid on the Islands, which wrought much havoc, also served to act as a «show of force» to the incumbent Arabs,whose vulnerability everywhere was now suspect.

The legends surrounding the victorious landing of Count Roger in Malta and the rapturous welcome he received from the Maltese are numerous; but they are misleading and probably unfounded. The name of Count Roger is still much revered in the Islands. It is said that he gave Malta her flag based on the Hautville colours. He is reputed to have re-Christianized the Maltese, established churches, re-appointed a bishop and even expelled the Muslims. It is always difficult to seperate truth from entrenched myths and legends, which were largely perpetrated by some historians at a later date for various well meant but ill judged reasons. When Malta was well and truly back in the Christian fold a few centuries later, the prevailing spirit of intolerence towards Islam within the uncompromising environment of Christian Europe, of which Malta formed part, encouraged the denigration of the Arab connection; as if it had never happened. Attempts were made to alienate and even eliminate it from Malta altogether.

The fact remains that although the Arabs surrendered easily to the Normans in Malta, they were not cast out or deported. True to their tradition, the Normans made no attempt to persecute them. Indeed they remained in control for more than another century. Count Roger never garrisoned the Islands. Arabic influence remained more or less unrestricted till about 1224, when the Muslims were finally expelled.

However, it has to be said that the appearance of the Normans in Malta in 1090 heralded a new era. It was perhaps symbolic of the steady erosion of the almost total Muslim hegemony over the Islands. In Malta, Count Roger introduced a new spirit of diversity, so characteristic of the Normans of the day. No doubt, this gave the Maltese some added confidence with which to shed the Arabic «strait jacket» they had now worn for so long, and which was possibly stifling their development and versatility. In that sense, he also opened the way towards the gradual re-entry of Malta into the Latin European mainstream, soon to be dominated by a militant Christianity best expressed in the spirit of the crusades.

The so-called Norman period in Malta which lasted till about 1194 was probably anything but Norman. Because Count Roger or his successors never attempted to colonize the Islands, they remained Arabic in essence. Indeed, there was a resurgence of Muslim power following the death of Count Roger in 1101. His son and successor, Roger II had to mount a second re-conquest in 1127. From this moment on, Malta started to reap the benefits of a general expansion of trade within the whole Mediterranean basin. This was largely conducted by the rival commercial and maritime empires of Pisa, Genoa and Venice. Within his realms, Roger II proved to be even more enlightened than his father. An Arabophile through and through, he even wore oriental dress. Under his patronage, Arab arts and crafts flourished in Sicily and, to a much lesser extent, in Malta. Regrettably, the Islands remained the poor relation and, whereas in Sicily the Normans left many art treasures and fine architecture, hardly anything of this kind exists in Malta.

However, we should at least be thankful for the discovery in Gozo of one of the most important and beautiful relics of this period. This is the marble tombstone of a young Muslim girl called Maimuna, who died in the year 569 of the Hegira or A.D. 1174. Carved in elegant Kufic letters, it bears the following evocative inscription.

«Ask thyself if there is anything everlasting, anything that can repel or cast a spell on death.
Alas, death has robbed me of my short life; neither my piety nor my modesty could save me from him.
I was industrious in my work, and all that I did is reckoned and remains.
Oh, thou who lookest upon this grave in which I am enclosed, dust has covered my eyelids and the corners of my eyes.
On my couch and in my abode there is nought but tears; and what will happen when my Creator comes to me?» [10]

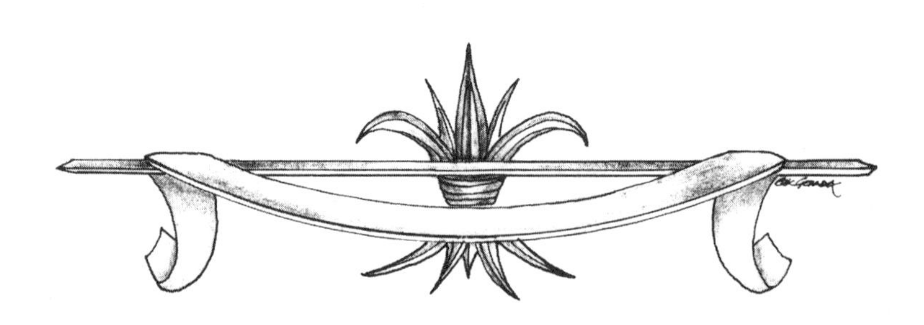

References
1. Luttrell: Op. cit. p. 25
2. Vella: Op. cit. p. 65
3. Luttrell: Ibid. p. 27
4. Bellanti: Op. cit.
5. Vella: Ibid. p. 67
6. Vella: Ibid. p. 66
7. *Blouet* Brian: The Story of Malta (Faber and Faber, London 1967) p. 41
8. Bradford: Op. cit. p. 320
9. Bradford: Op. cit.
10. As quoted in Bradford: Ibid. p. 333

MEDIEVAL MALTA

With the Norman re-conquest of Malta in 1127, King Roger II immediately recognised the growing importance of the Maltese Islands to Christendom within the changing Mediterranean.

With the Norman fleets now dominating the Central Mediterranean and with those of the Byzantines in control of the East, the region had entered into an era of relative stability. Muslim naval power had seriously declined and the Mediterranean was now secure from any serious Arab threat. The important sea lanes between Europe and Africa and the Levant were thus reopened to Christian shipping. There was suddenly a marked increase in maritime activity on the high seas which was all leading to a greater expansion of trade in the area.

In Malta, Muslim power and influence were still strong. These now had to be curbed. For this reason, the Normans quickly installed their own «Emir» and his men supplanted the Arab administrators. There is no evidence of any persecution of Muslims at this stage. However, it is clear that Malta was now moving more and more into the Latin and Christian sphere of influence. At the same time, there is equally no evidence of any substantial Latin colonisation as yet. Certainly, the Normans must have kept a small garrison and maintained the main fortifications at Mdina, in the Grand Harbour and in the castle on Gozo. Because Malta was still nominally Muslim and non-Latin speaking, the Normans did encourage the religious orders, especially the Benedictines, to establish a Christian community in the Islands. However, these attempts were unsuccessful and the Benedictines did not arrive till much later on.

At the spearhead of the new maritime advance in the Mediterranean were the Italian ports of Pisa, Genoa and Venice. As east met west, their fleets were to be seen everywhere. Christian Europe was suddenly alive. She had been galvanized into action in 1091 when the first rallying call was sounded in Clermont by Pope Urban II. From feudal Europe, the great armies of the first crusades were making their way to the East. Their declared and «holy» duty was to drive out the infidel Turks and Saracens out of the Christian «Holy Places» in Jerusalem. The wars of religion had been unleashed and they were to last for over two centuries. In their wake, the great maritime powers of Pisa, Genoa and Venice rose to fame and fortune. Inevitably, with the Christian barons, they became heavily involved in the logistics of supply and movement of armies of men to their battle destinations in the East. With every renewed crusade, their power and influence grew.

The Normans were more or less «allied» to the Pisans and the Genoese. As a result, Malta too was now benefitting from the increase in maritime activity around her shores. Again, she was playing her traditional role; acting as a staging post between Europe and Africa. Along with Sicily, the Genoese were using her as an operational base for their markets, now stretching far and wide into the Levant. It is possible that Malta was also acting as a spring board for pirate attacks on the increased merchant marine in the region. No doubt, many Maltese were recruited into the Royal Fleets. It is likely that they took an active part in the successful raid on the Tunisian pirate base at Djerba in 1135.

Malta's ports and safe anchorages were now in regular use. Slowly the Maltese were being drawn out of their previous relative isolation and introduced to the more cosmopolitan world of the Normans, the Genoese and the Pisans.

It is not known how Malta fared during the series of political struggles which rocked the kingdom of Sicily, following the death of King Roger II in 1154. His successor, William I, known as William the Bad, was a «playboy» king. Largely uninterested in state affairs, he delegated most of his duties to his ministers. Through them the chief beneficiaries were the Genoese, to whom many substantial privileges were granted. William was succeeded by his young son. Until he came of age, the kingdom was ruled briefly by Margaret of Aragon. When William II came to the throne, he further pursued his mother's policies, which were based on the closest of relations with the Popes and the established church. Perhaps for this reason, he went down in history as William «the good». It is about this time that mention is made of John, Bishop of Palermo and Malta. Although no evidence exists of his presence on the Islands, he was certainly active in Palermo during the period 1168-1212.

When William the Good died childless in 1189, a dispute arose over his successor. The rightful heir was the daughter of Roger I, Constance, who was married to Henry VI, son of the Emperor of Germany, Frederick Barbarossa. However, the Pope had other designs. Fearing the penetration of the Germans in Sicily and Southern Italy, the Church threw its support on to Tancred, an illegitimate son of Norman descent. Indeed, in 1190, Tancred was crowned King of Sicily. His reign did not last long since Henry VI had now quickly allied himself to the powerful Genoese. Through a series of intrigues within Tancred's court, Henry acquired Sicily in 1194.

In the meantime, Tancred had elevated a renowned Genoese pirate, by the name of Margarito of Brindisi, to the exalted position of Royal Admiral of Sicily and Count of Malta. This was a new development for Malta. For the first time the Islands were thus seperated from the Royal demanium. Genoese influence in Sicily was clearly very high since another Genoese corsair, William Grasso succeeded Margarito soon after. Largely because of Grasso's unpopularity in Malta, Constance of Sicily, by now restored to the throne, decided also to restore the Island to the crown lands in 1198. A year later she died.

The heir and successor to the Kingdom of Sicily was now Frederick II, also Emperor of Germany. With Sicily, Malta now entered into the era of Hohenstaufen rule. However, until Frederick came of age in 1220, the Kingdom of Sicily went through yet another unsettled period.

In Malta, Grasso's son-in-law, Henry Pistore, became Count of Malta in 1203. Also Genoese, Pistore was a colourful figure and most ambitious. It is said that in 1205, with the assistance of Maltese sailors he set off with two galleys, and after commandeering another two Venetian galleys, he attacked Tripoli in Syria. So successful was his expedition, that he and the Maltese were welcomed back to Malta as national heroes. Pistore's use of Malta as an independent base for his various incursions must have raised Malta's status in the Mediterranean and probably diversified her sources of income. Genoese influence in Malta was now at its highest ebb. It was from Malta that a strong Genoese fleet set sail to successfully attack Syracuse with great success in 1205 after the Pisans had seized it. In 1210, Pistore mounted an expedition to Crete where, for a brief period, he set up a kingdom. In 1221, he was appointed Admiral of Sicily by Frederick II, but he soon fell out with the Emperor. The great era of Frederick II, «the wonder of the world» had now commenced. From this moment on, it is his name that totally dominates the scene in Italy, Sicily and Malta.

Frederick II, Ruler of Sicily and Germany, Holy Roman Emperor and King of Jerusalem was an ambitious monarch, determined to restore in his dominions the ideals and greatness of the reign of his illustrious Norman grandfather, Roger II. During his lifetime, he combined the Norman tradition of diversity and patronization of Islamic arts and sciences with a great independence of mind and spirit. Demanding total loyalty to him in affairs of state, he was an absolute monarch exercizing firm control over his realms. A forerunner of the concept of the separation of Church and State, he immediately came into conflict with the Pope, whose revenues and influence he sought to curb. In his quest for total allegiance, Frederick also alienated himself from many feudal barons. His harsh treatment of the Arabs was curious since he was basically a great Arabophile. As the symbol of his kingdom, he chose the falcon, the «king of birds» much revered in the Orient. «He also kept a harem and his court abounded with men of learning and dancing girls from the Orient»[1]. He even wore oriental clothes. Despite this Arab trait, Frederick sought to dislodge rebellious Muslims from his domains. He immediately quashed an Arab uprising in Sicily which also had the tacit support of the Muslims in Malta. This resulted in a purge of all Arabs and by 1224, he had them expelled completely from Sicily and Malta. He then colonised the «vacuum» thus created in Malta, by bringing in the Latin population from Celano, who themselves had been expelled from their home by way of a reprisal for their «treason» when they supported their feudal barons against him.

Honouring Constance's pledge, Frederick II restored Malta to the Royal demanium. Through his administrators and his Governor Paolino, he kept a careful record of the Royal lands which included the three castles at Mdina, Castrum Maris at Birgu and that on Gozo. By 1239, Malta was thus again united with Sicily. Curiously, Paolino also had to account for the eight camels he sent to Frederick, and for another three which were kept in Malta for breeding purposes. In 1240, Frederick sent 18 falconiers to Malta to report on the number of falcons being caught in Malta, then famous as a breeding sanctuary for these birds of prey.

There is much controversy over the population statistics contained in the report on the Maltese Islands submitted to Frederick by his agent Gilibertus in 1241. Here it was stated that out of a total of 1119 families in Malta and Gozo, as many as 836 were Muslim and that the majority of Christian families were Gozitan. However, no real conclusions can be safely drawn from it since the basis of what constituted a «family» especially in terms of revenue received from the Royal lands, was not defined and therefore subject to dispute itself. It is likely that the majority of Muslim Maltese immediately «re-converted» to Christianity to avoid expulsion. Had the total Muslim community, which comprised the bulk of the population of the Islands been expelled, the Arabo-Berber element of the Maltese language would have faded forever.

In 1250, that «great lover of Oriental luxury»[2] the Holy Roman Emperor Frederick II died excomunicate and was buried in the Cathedral at Palermo. His epitaph reads:

«If it were ever possible to achieve immortality through honesty, wisdom, power, wealth and nobility of descent, then Frederick who lies buried here, would never have died».[3]

The decline of the Hohenstaufen dynasty was rapid. With the great Frederick out of the way, the many enemies he made during his lifetime quickly moved in to redress the balance. Chief among these were the Popes; Innocent IV and later Alexander IV. The Church was as determined as ever in her resolve to rid Italy and Sicily of the Germans. Through plot and counterplot, and in direct connivance with any party willing to assist in the overthrow, the Popes were largely unsuccessful at first, since Frederick's son Conrad succeeded him to the throne. However, it was Manfred, his other son, who as Prince of Taranto presided over the Royal domains in Italy and Sicily. When Conrad died in 1258, it was indeed Manfred who was crowned King of Sicily.

Manfred made several attempts to patch things up with the Pope. Despite his frequent appeasements in favour of the church, Pope Urban IV remained unmoved in his resolve to overthrow him. The last straw came when Manfred's daughter, Constance married Peter, the son of James of Aragon. The Pope's only hope now lay with Charles of Anjou in France whom he regarded as the «defender of the Church». Charles was clearly his favourite to rule over Sicily and thus safeguard the Church's interests in the kingdom. In this maze of intrigues which characterized these violent times, Manfred was facing enormous difficulties everywhere. In order to regain some of their old privileges and restore their influence in Sicily and Malta, the Genoese now masterminded an uprising among the population in Malta. They based their claims on certain rights over Malta which Nicolosio, son of the colourful and popular Henry Pistore, was now demanding. In the hope of appeasing the Genoese, Manfred agreed to most of Nicolosio's claims in 1257. However, he kept the main fortifications on the Islands under his control.

In the meantime, the Popes were slowly closing in on Manfred. Pope Clement IV threw all his support, both spiritual and temporal, in favour of Charles of Anjou. He was determined that neither the Aragonese, through Manfred's daughter Constance, nor the young Conradine, son of the dead Conrad should ever reign in Sicily. The matter had to be resolved once and for all. In 1266, Pope Clement finally achieved his objective. After proclaiming Charles of Anjou as King of Sicily, he formally crowned him. It was natural that Manfred should now immediately resort to arms to regain his kingdom. Regretably, he was no match for the Angevins and in the battle at Benevento, Manfred was defeated and killed. All eyes were now on Conradine. In 1267, he conspired for the support of the Pisans in return for which he undertook to hand Malta to them. His plan failed. At Tagliacozzo, Conradine was captured and in 1268, he was executed in Naples. Within the space of two years of Pope Clement's proclamation, Charles of Anjou had emerged totally victorious and in firm control over Sicily and Malta.

Although the period of Angevin rule over Malta was shortlived, it is from this point onwards that we have evidence of the Islands' rapid shift into the European and Latin schemes of Government and administration. The Angevins quickly imposed very firm and tight control over the Islands. Malta now had a Governor or «Castellan» who commanded a garrison of French troops. His duties also included the protection of shipping in support of the castle on Gozo. Detailed inventories of the Royal lands and other possessions were drawn up. Castrum Maris in the Grand harbour was now clearly being prepared to act as a well fortified and equipped centre of resistance to fight potential aggressors.

Malta and Gozo were now also enjoying a measure of representation in the government of the Islands through their local councils or «Universita». During this time, a notary public to serve both islands was also appointed. However, the Maltese were far from happy. The burden of the rather high royal taxation imposed on them made life difficult.

It was also during this time that the Church was slowly consolidating its position among the laity. From about 1270, it would appear that an organized diocese was in existence, although still as a suffragan to the Bishopric at Palermo. It is said that in 1272, a Franciscan, by the name of Jacobus de Malte, was Bishop of Malta[4]. French rule in Malta was clearly unpopular but it was possibly more benevolent than that in neighbouring Sicily. Increasingly restive under the Angevins, the Sicilians were already plotting to restore the island to Constance and Peter of Aragon, the rightful heirs to the crown in Sicily. The Aragonese immediately sought to recruit the support of the Genoese and the Venetians to fulfil their objective. No doubt, Malta became involved in this intrigue, for in 1272, the Angevins seized two Genoese ships in the Grand harbour. In revenge, the Genoese sacked Gozo in 1274. Things came to a head in 1282 with the Sicilian uprising against the French. This revolution, now refered to as the «Sicilian Vespers» led to the bloody massacre of the French in Palermo. Taking swift advantage, the Aragonese immediately installed Peter of Aragon as ruler of Sicily.

Malta was still held by the Angevins and it was here that their fleet regrouped. The Angevins now strengthened the fortifications of Castrum Maris in the Grand harbour and prepared themselves for a counter assault on Sicily. From Marseilles, reinforcements arrived in Malta and a large Angevin fleet was now assembled ready for the coming battle. A head-on collision had to take place sooner or later. The Aragonese were not only determined never to give up Sicily but, in line with their strategic designs for the central Mediterranean, Malta too had to be conquered. In 1283, the Aragonese fleet appeared off Malta for the decisive battle. The Grand harbour witnessed the first major battle to be fought in its waters. The Angevin fleet was routed and their Admiral, allegedly an Englishman named William Corner, was killed in the battle. The Maltese, clearly on the side of the Aragonese welcomed their Admiral with much jubiliation. Under their Field Commander, Manfred Lancia, the Aragonese then went on and captured Castrum Maris. Although officially already forming part of the Aragonese Kingdom of Sicily since 1282, it was through this decisive battle fought in Malta, that the Islands' fate with the Aragonese was finally sealed. The period of Aragonese rule over Malta had commenced and it was to last for a long time.

Despite this initial jubiliation on the part of the Maltese, disenchantment with their new rulers quickly set in. From the start, it was clear to them that they were to become mere pawns in the power struggles then taking place in the Mediterranean. Malta's destiny under the Aragonese was to be at the mercy of one feudal overlord after another. For this reason, during the whole of this generally unhappy period, we will see the Maltese continually petitioning the crown to restore the Islands to the Royal demanium. They also knew that ultimately Sicily was to hold the key to Malta's prosperity or misery.

The Mediterranean was again changing rapidly. As the new drama unfolded, it centred largely on developments in Sicily. The chief actors were the Aragonese, now intent on expanding their empire and extending their trade routes. Next came their enemies, the Angevins, smarting from their recent defeats in Malta and Sicily, but still holding on to Naples. Lurking in the wings, and always at the ready for any prize that might come their way, were the powerful fleets of Genoa and Venice. Orchestrating the scene, from an overall position of strength and influence stood the guardians of militant Christianity, the Popes in Rome, themselves reeling from their many reverses in «Outremer» as the Latin kingdoms of the East were then referred to. Indeed, from this region was emerging the next versatile actor of them all - the Ottoman Turk. Islam was again regrouping; and even in the central Mediterranean, the Hafsids from Tunisia were already making the odd menacing appearance here and there.

The death of Peter The Great in 1285 caused many divisions and dissensions within the Aragonese court and in the Empire, which now included Catalan and Valencia, the Kingdom of Sicily, Malta and many other commercial and trading interests in Africa and other islands in the Mediterranean. Peter's son, Alfonso, now became King of Aragon; while Sicily was bequeathed to his other son, Jaime. As a result, many divided loyalties with as many betrayals were thus caused. When Alfonso died five years later, Jaime succeeded him to the throne of Aragon. In the meantime, the Angevins in Naples, with the support of the Pope, were still bent on the reconquest of Sicily and Malta. They very nearly succeeded in realising their aims when in 1295, Jaime made a secret deal with Pope Boneface VIII for the barter of Sicily and Malta in favour of the Angevins, in exchange for Sardinia and Corsica. However, the Sicilians were quicker on the mark. They immediately repudiated Jaime and proclaimed his other brother, Frederick III as King of Sicily.

Malta too got caught up in the feud which now arose between Jaime, King of Aragon and Frederick who controlled Sicily. The Aragonese fleet, loyal to Jaime made a savage attack on the

Islands in 1296. This was only a foretaste of the misfortunes and exploitation which were to follow. Always anxious to maintain the loyalty of their feudal barons, on whose material support they depended, the successive kings of Sicily now continually granted Malta as a fief for favours or services rendered to the Crown. In 1350, we find the Maltese petitioning King Ludovic for Malta to be restored to the Royal lands. However, no sooner had an agreement been reached, when Jeanne of Anjou, Queen of Naples and temporarily in possession of Palermo, ceded Malta to the Florentine, Nicolo Acciaiuoli. When the Neopolitans were finally expelled from Sicily, Maltese hopes of a return to the Crown were again dashed. In 1360, Frederick IV installed Guido Ventimiglia as Count of Malta. It was now clear that the Sicilian barons were increasingly dominating the political scene. In order to ensure their support and dubious loyalty, the King was perhaps too liberal in granting royal lands in return for faithful service. As the power and influence of the barons increased, they became even more avaricious and rebellious. Malta soon passed through the hands of a succession of such Sicilian magnates. The names of Chiaramonte, Admiral of Sicily and of Giacomo Pellegrino, Captain of Malta, feature prominently in these troubled times.

The general uncertainty and apathy which hung over the Maltese Islands was leading to much local unrest among the population. They were tired of the additional burdens in taxation which every new despot now imposed on them. Increasingly, they now turned to piracy for additional revenue with which they could keep their heads above water. Regrettably, their attacks on shipping may have been somewhat indiscriminate since these often led to reprisals against Malta. Indeed, to settle old scores, the Genoese attacked the Island in 1371. This is turn provoked additional discontent among the inhabitants. It was also clear that the rebellious pretensions of the Royal Captain of Malta did not go unnoticed by the Crown. Possibly for this reason, Frederick IV visited Malta in 1372. He immediately sacked Pellegrino and rewarded in turn those who, like Lancia Gatto, had supported the Royalist cause. But disloyalty and betrayal seemed to be the order of the day in these times, since yet another royal protegé Frederigo d'Aragona was ousted from Malta in 1376.

With the death of Frederick IV in 1377, the Kingdom and Malta with it went through yet another period of confusion, uncertainty and anarchy. Again, only the Sicilian magnates were to benefit from it. Again, the name of Chiaramonte suddenly looms large on the local scene. His successful raid on the island of Djerba in 1388, which he possibly led from Malta provoked a reprisal on Gozo by the Hafsids a year later.

Malta now remained at the mercy of the powerful Sicilian magnates, like the Alagonas and the Moncadas. It was not till 1397, that the local councils of Malta and Gozo - the «Universitates» - made a strong petition to the Crown for the Islands to be restored to direct rule by the King. They also complained about the stiff import duties then being levied on basic foodstuffs and animals from Sicily. These taxes were most objectionable since Malta was, after all, part of the Kingdom of Sicily. The Maltese were largely successful in achieving their main objective. However, they were now obliged to pay to the Crown 5% of the booty taken from their pirate raids on shipping. In addition, they had to make a contribution towards the maintenance of the main fortifications of the Islands, which were now in a bad state of repair.

The lethargic pace of events did not bring much solace to the Maltese in their struggle for survival. The situation was made even worse with the growing schism now unfolding within the Church. The bitter struggle between the Roman Pope and the other pretender in Avignon, whom the Aragonese supported, also had a divisive effect in Malta. This was taking place at a crucial time for Christianity in Malta, when the great religious orders were finally being persuaded to set up communities on the Islands. By 1372, both the Benedictines and the Franciscans had arrived. With their appearance, Christianity now took firm root among the population and flourished in a quite unprecedented manner.

In 1412, Ferdinand de Antequera was elected King of Aragon, Castille and Sicily, the first Castillian to ever occupy the throne. In 1416, the Maltese petitioned his son Juan, Viceroy of

Sicily to implement a number of reforms in Malta. The Island's security system had deteriorated badly as a result of the many malpractices then taking place in the rosters for guard duties. For this reason, the Islands' main fortifications like Castrum Maris at Birgu had become particularly vulnerable. Juan agreed to effect improvements. He also showed much sympathy with the sad plight of the Maltese who were now the subject of incessant attacks from Saracen pirates. Recognizing the poverty now prevalent among the population, Juan also agreed to ease the import duties on provisions from Sicily. Regrettably, most of the reforms hardly saw the light of day since in the same year, Alfonso V succeeded to the throne. Known as «the magnanimous», Alfonso did not quite live up to his name during the initial period of his rule over Malta. An ambitious King, he was determined to expand Spanish interests and influence over the length and breath of the Mediterranean. To finance his adventures, he resorted to the «pawning» of his island possessions to willing and loyal Viceroys. They in turn, enjoyed all incomes and dues from the Royal lands. Additional sums, through other fiscal measures were then extorted from the inhabitants. Repeated Maltese petitions for the improvement of the Islands' defences remained unheeded. The more recent request, for which money was raised, to build a castle on the island of Comino, which had now become a pirate's den, serving as a springboard for their attacks on Gozo and Malta, was also shelved.

In January 1421, King Alfonso granted the Maltese Islands and all revenues from them to Don Antonio Cardona in exchange for a loan of 30,000 gold florins of Aragon. However, in March of that year, we now find Cardona «transfering his rights over Malta and Gozo to Don Gonsalvo de Monroy».[5] The Maltese clearly disagreed with this new arrangement. It took another five long years until they finally rebelled. In 1426, they pillaged Monroy's house in Mdina and laid seige on the castle at Birgu where his wife was esconced. Malta was in Maltese hands for a brief period.

Empowered with the full authority of the Maltese «Universita», a delegation was immediately sent to Sicily where the Maltese demanded the right to «buy out» Monroy for 30,000 florins. They also insisted on radical reforms and for certain guaranteed rights to be granted to them; particularly those relating to the appointment of Maltese to key positions in the Islands' administration. In addition, it was made clear that the Islands must never be ceded again by the Crown. With some reluctance Alfonso agreed to the new reforms. In 1428, he formally ratified Malta's rights under the Crown in a Royal Charter.

The jubiliation of the Maltese at this turn of events was somewhat dampened when a year later, the Hafsids from Tunisia laid seige to Malta. The recurrent problem of the lack of adequate defence arrangements was all too clear. The Hafsid invasion, with a force of about 18,000 men can now be regarded as Malta's first serious seige, from which the Maltese, although severely mauled and crippled, came out victorious. Although the Hafsids caused havoc everywhere, killing many Maltese and enslaving a good part of the population, they did not succeed in capturing the Islands. Nevertheless, the repercussions of this attack were far reaching. The economy of the Islands was to remain in ruins for many years. Many villages, some almost totally depopulated during the seige, were now completely abandoned. For years to come, the Maltese struggled to raise ransoms in order to free their enslaved brothers from the Hafsids. Attempts were also made to fill the population «vacuum» and thus hopefully revive the economy. Foreigners were now openly encouraged to settle on the islands. The Maltese again took to piracy with renewed vigour. This flourished to such an extent that in 1440, a ban on the recruitment of Maltese sailors by foreigners was imposed. By 1460, a Maltese pirate, Michele de Malta had became famous for his exploits in the Aegean. Generally, times were lean in Malta and there was much hardship. The rampant malpractices and corruption in government, which continued unabated despite repeated protests by the Maltese to each new Viceroy, only contributed to the deteriorating situation of despair and apathy.

In 1458, Alphoso was succeeded to the throne by his brother Juan II. His reign was marked by the bitter struggles which took place in Sicily and in Catalan. It was not till 1479 that some stability was restored. Through the marriage of Juan's son and successor, Ferdinand II to Isabelle of Castille, Spain was politically re-united once more. A new era was in the making.

By this time the Ottoman Turks were already making their mark in the Mediterranean. They had made impressive gains in North Africa and their fleets were now well poised to menace Christian shipping and other possessions in the Western Mediterranean. In 1487, they landed at Marsaxlokk in Malta and marching into Birgu, ransacked the city capturing 80 Maltese. It was this sad event that finally prompted Ferdinand to take a more active interest in the Islands' defences. This must have boosted morale and general confidence in Malta, since it is at about this time that population figures started to increase. The Maltese, always willing recruits into the fleets of their masters took an active part in Ferdinand's campaign to capture Tripoli in North Africa in 1510. Malta was now again acquiring a significance of her own, as she lay poised between the seat of Spanish traditional interests in Sicily and her newly captured «prize» in North Africa. With the now swift penetration into the Mediterranean of the Turks, the importance of Malta to European Christendom was becoming increasingly clear. In 1526, the Turks struck again, this time landing at Salini. From here, they then raided the village of Mosta killing many Maltese and enslaving about 400 villagers.

In the meantime, important developments were also taking place in Europe. Spain and the Hapsburgs of Austria came together when Joanna, the daughter of Ferdinand II married Philip, Archduke of Austria and Duke of Burguandy. In 1518 the Hapsburg dynasty was consolidated when their son Charles V became Holy Roman Emperor and through the intercession of Pope Clement VIII, he finally decided to grant Malta, Gozo and Tripoli in North Africa to the now «homeless» Order of St John in 1530. At a time when Ottoman Turkey all but ruled the waves in the Mediterranean, challenging Christendom to its very limits, this enfoeffment of the Maltese Islands to the Knights of St John seemed a sensible and inexpensive way of securing their defence. Concurrently, the Knights were obliged to maintain the defences of Tripoli in North Africa. The proven enemies of Islam were thus re-introduced into the arena of battle against the Infidel. Their task lay in sweeping the sea lanes of the anarchy that existed as a result of their domination by the Turkish Fleets. This way, some protection to the southern flank of Christendom would be achieved.

The stage was again being set for one of the greatest dramas ever to be played out in the Mediterranean. The chief actors were now the opposed champions of Christendom and those of Islam; both seasoned veterans of battle, of seige and counterseige. For the annual rent of one falcon, which the Knights of St John were obliged to pay Charles V for their possession of Malta, the Holy Roman Emperor had struck the bargain of the century. He had saved Christian Europe from Islam and initiated the ultimate decline of Ottoman Turkey.

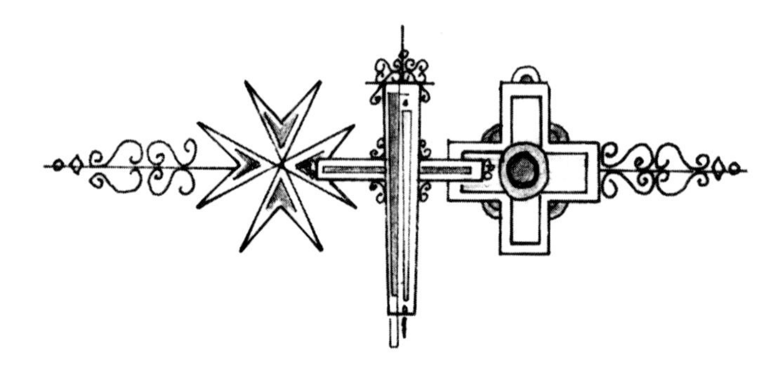

1. Anthony *Nutting* : The Arabs (Clarkson N. Potter, Inc., New York 1964 and Mentor Books, New American Library, New York. Fourth printing) p. 154 (Mentor Book Ed.)
2. Nutting: Ibid. p. 193
3. Vella: Op. Cit. p. 95 Translated into English from Maltese by the author of this text.
4. Luttrell: Op. Cit. p. 61
5. Godfrey *Wettinger* : in "The Pawning of Malta to Monroy' in Melita Historica, Vol III No. 3 (The Historical Society, Malta 1978) pp. 265 - 283.

THE KNIGHTS OF ST JOHN

OUTREMER

The origins of the Sovereign Military Hospitaller Order of Saint John of Jerusalem, of Rhodes and of Malta are irrevocably linked to the spread of militant Christianity throughout Europe. As the established Church grew in strength and stature, it was natural that its members should take a more active interest in the birthplace of Christianity and in those areas where its Founder preached the new ethic in the exercise of His divine mission to redeem mankind. The attention of fervent Christians was thus drawn to the Holy Lands which centred around the old city of Jerusalem.

In AD 637, this great city had fallen to the Arab armies of the «new» religion of Islam and the whole surrounding region had remained under their control for many centuries. However, Christians were never persecuted in the conquered lands. Pilgrims were free to visit the Christian holy places and shrines in Jerusalem and elsewhere. From Europe they came, either via the overland routes across Asia Minor through Byzantine territory or by sea, across the Mediterranean in the merchant ships of Venice and Amalfi. These Italian cities had, over the years, greatly expanded their trade with the east. There, they had set up a network of commercial outposts. Through them, they could also service the requirements of pilgrims on arrival at their sacred destination. In Jerusalem, some Benedictine monks, themselves originally from Amalfi, had also established a hostel where weary pilgrims could find shelter and succour. They would be the forerunners of the Knights Hospitallers.

Towards the close of the Tenth Century, this pattern of relative order and stability was suddenly disturbed. Sweeping in from Asia Minor, the primitive Seljuk Turks were quickly replacing their former Arab masters in the major cities. Ready converts to Islam, these warring tribesmen became increasingly militant and soon they were menacing not only the pilgrim land routes but also challenging the stronghold of Christianity in the east, Byzantium.

The fall of Jerusalem to the Turks in 1071, soon to be followed by that of Syria, finally provoked the Byzantines to appeal to their brothers in the Christian west for support against the new common enemy of Christendom, now threatening all access to the Holy Places in Jerusalem. Despite the schism that existed between the Roman Church and that of Constantinople, their call did not fall on deaf ears. In 1091, Pope Urban II galvanized Christian Europe into action. By 1096, about 20,000 Christian fanatics descended upon Asia Minor. However, they were nothing more than a disorderly rabble. Their faith was not enough to save them. In no time, they were routed and massacred by the waiting Turks.

Christian Europe was outraged at this initital disaster. With a vengeance, Christendom now mobilized itself and embarked on an organized campaign to redress the balance. In charge were the Normans, those well seasoned veterans of many a seige. In 1097, a multi-national force of about 150,000 men at arms besieged the major cities along their route to Jerusalem. By 1099, they had breached the walls of the great city itself. They poured into Jerusalem, systematically butchering all Muslims and Jews, including women and children. Hailed as a great victory for Christendom, the bloody fall of Jerusalem to the First Crusaders was never to be forgotten or forgiven by Islam. For the next five centuries, the wars of religion between the now sworn enemies of Christendom and Islam were to rage unabated.

In charge of the Benedictine hostel at the time of the seige of Jerusalem was a monk known as Brother Gerard. Under his care, the hostel had acquired facilities for the treatment of sick pilgrims. The monks were then known as the Hospitallers of St John of Jerusalem, after the patron Saint of their convent St John the Baptist.

It is likely that, along with other Christians, Peter Gerard and his followers were expelled from Jerusalem before the gruelling six-week seige commenced in earnest in 1099. Their medical services must have been most welcome to the many wounded crusaders during the course of the great battle. No wonder then, that the hostel with its little hospital was amply rewarded with gifts of money by grateful Christian noblemen who survived the seige. Soon, more generous donations and even land was bequeathed to it. This greatly assisted towards its expansion. During Gerard's lifetime, sister-houses were established along most of the pilgrimage routes in Europe and elsewhere. Gerard, «the most humble man in the east and the servant of the poor»[1] never deviated from his original mission. Forseeing the inherent dangers of this sudden rise to fame and fortune of his convent of monks, he laid the basic foundations for its evolution into an established monastic Order. Indeed, by 1113 Pope Pascal II had acknowledged the Hospitallers as a religious Order. Bound to the Augustinian rules of Chastity, Poverty and Obedience, the care of the sick and poor remained its chief concern.

It was Gerard's successor, Raymond de Puy who extended the role of the Order to include the protection of pilgrims on their way to Jerusalem. A military arm was thus created. The result of this development was an increasing sense of militancy against Islam within the Order. In an age which saw the formation of many military orders of chivalry to fight this or that cause, the Order of St John emulated the other Orders like the Knights Templars and adopted many of their feudal institutions. The exercise in arms in defence of the Christian cause soon outstripped the original mission to tend to the sick.

Now, only noblemen were admitted to full knighthood. By the middle of the 12th century, the Order was headed by a Grand Master. Its wealth, power and influence increased considerably. By 1168, among its many castle possessions was the famous Krak des Chevaliers. The Order was soon taking an active part in the major fields of battle against the infidel. Here, the flag with the white cross on a red background of the Knights of St John became a common sight alongside those of the other Christian armies. But it was the four arms bearing eight points of the cross in white against a black background, which the Knights wore on their person (and now known as the Maltese Cross) which perhaps was to become more famous. The arms of the cross represented the four Christian virtues of prudence, justice, fortitude and temperance; the eight points were the beatitudes.

The first serious setback to «Outremer» - the Latin kingdoms of the East - came with the fall of Jerusalem to the unified armies of Islam under the great Saladin in 1187. However, their decline had already been heralded by the failures of the Second Crusade from 1148 onwards. The Third Crusade, which was mounted after the loss of Jerusalem, was perhaps more successful. At least most of the coastal areas, including the city of Acre were recaptured. But Jerusalem was lost forever. Bitter divisions within Christendom now arose. When the Fourth Crusade was assembled, ostensibly aimed at Muslim power in Egypt, the crusaders betrayed themselves and the Christian cause when, along their route, they sacked Constantinople-that old eastern bulwark of Christianity itself. They then carved up the Byzantine Empire among themselves and quickly forgot their holy mission in Egypt. The Christian cause in the East was clearly evaporating. In 1244, the

Grand Master of the Order of St John was himself captured when Gaza fell. The Fifth Crusade, led by St Louis of France ended in disaster. He too, was captured in 1249.

The retreat from the Holy Land by the Order really got under way when their castle, Krak des Chevaliers fell to Babyers, Sultan of Egypt and Damascus in 1271. Fortress after fortress soon succumbed to the armies of Islam, now determined to rid the region of all vestige of Christianity and to crush their deadly enemies; the Military Orders of St John, the Knights Templars and the Teutonic Knights. The Order of St John had now established itself in the old harbour city of Acre in Palestine. Here, they prepared themselves for the inevitable siege on the last Christian stronghold in the Holy Land.

It came in 1291. The Turkish Mamelukes stormed the fortifications and defences of the city at every point. After a month - long seige, the vengeance of Islam was fulfilled. Acre was totally destroyed. Shortage of sea-going vessels in the harbour prevented the complete evacuation of the surviving Christian population and soldiery. Those who remained were either slaughtered or enslaved. As refugees poured into Cyprus, it was clear that all was lost. The dream of Outremer was over.

From now on, the activities of the Knights of St John would be confined to the Mediterranean and its islands. As any sailor will tell you, only good sailors survive in the Mediterranean. The soldier-monks thus had to acquire a new role.

There was, however, a yet more valid reason in favour of the Knights' taking to the sea in the way they did. Unable to fight Islam on land the Order had to be seen to be continuing the battle at sea, at least by harassing the enemy in the Aegean. It was therefore at Limassol in Cyprus, where the Knights were now based for a period close on to 20 years that a strong naval arm was first formed. In Cyprus, the Order of St John was largely unwelcome. Their reputation as the favoured military order of the Popes, combined with the considerable influence which they wielded in the courts of the European Kingdoms, were clearly resented by the Cypriots, anxious as they were to preserve sovereignty over their kingdom. Indeed, it was sovereignty over any future «home» that the Order was aspiring to.

For this reason, the Knights now looked to the island of Rhodes, itself surrounded by the other little islands of the Aegean. With the assistance of the ubiquitous Genoese, they attacked the Island in 1307. For two years the Rhodians resisted the invasion. Finally, they surrendered.

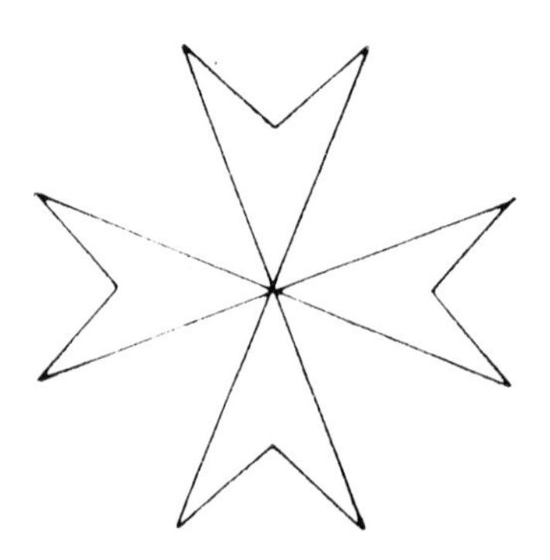

1. Ernle Bradford : From Gerard's epitaph quoted from The Shield and the Sword : The Knights of St John. (Hodder and (Stoughton, London 1972) p 25

RHODES

The Military Hospitaller Order of St John of Jerusalem, homeless since the fall of Acre, had now acchieved its objective. Secure and sovereign over its new island home in Rhodes, it could now pursue its activities unhindered. Owing allegiance only to the Pope in Rome, it could at last commence necessary reorganization from within and slowly rebuild its material power and strength in depth. First, however, in keeping with its traditional holy mission to care for the sick, it would found a hospital in Rhodes.

Over the next two centuries, the Order of St John was to discipline and dedicate itself and thus emerge as the thoroughly efficient force that it again became in the east. With the considerable revenue from its lands and priories in Europe, it could finance the building of the fortificatins on Rhodes and those on the perimeter islands - its first line of defence. With its well equipped navy and its superior fire-power, it could guard and dominate the major sea lanes in the Aegean. Nor would it hesitate to engage Islam on land and conquer territory in Asia Minor as it did in 1345 when the Turks were driven out of Smyrna. With its Christian allies, it would assist in the capture (and massacre) of Alexandria in 1365. Over a period of 200 years, it would generally become the acclaimed chief outpost of Latin European Christendom in a menaced region where Islam too was not only effectively regrouping itself but also commencing its penetration of Eastern Europe and mainland Greece.

The fall of Constantinople to the Turks in 1453 heralded the militant advance of the Muslims. Inspired by their great leader Mehmet, the Turks launched an attack on Rhodes in 1480. Initially, their siege was successful. They even breached the fortifications of the city. However, they failed to take it. A counter-attack by the Knights finally drove them out. With this victory, the Order of St John was hailed in Europe as the invincible champion of Christendom. For Islam, it meant temporary decline; a time to reorganize.

In times of adversity, Islam had always shown a remarkable capacity to throw up great leaders to redress the balance and seek revenge. As Europe basked in the splendours of the Renaissance for the next forty years and as the Knights of St John prospered in its wake, that leader emerged in the person of Suleiman The Magnificent, Sultan of the Ottoman Turks, «Allah's deputy on Earth, Lord of the Lords of this World, Possesor of Men's Necks, King of Believers and Unbelievers; King of Kings, Emperor of the East and West».[1] His was a singular resolve: the expulsion of the Order of St John out of Rhodes - «that abode of the Sons of Satan».[2]

In 1522, Suleiman descended on Rhodes with an army of 200,000 men and some 700 ships. For six months, the Turks laid seige against the well fortified bastions and escarpments of Rhodes. Slowly, the defences started to cave in; but the Knights held on. Both sides were exhausted, especially the Rhodians who now seemed in favour of accepting the terms of surrender which the Turks were presenting. With great reluctance, Philippe Villiers de L'Isle Adam, Grand Master of the Order was compelled to surrender the Island to Suleiman in person. After complementing the Knights for their brave stand during the siege, Suleiman offered L'Isle Adam and the Order safe passage out of Rhodes and even material assistance in its evacuation.

The Sovereign Military and Hospitaller Order of St John of Jerusalem and of Rhodes was suddenly again homeless.

The expulsion of the Knights from Rhodes came as a rude shock to them. It was sad that the crash should come just when the Order had perfected itself and when it was at last reaping the benefits of a sound organization, now running at peak efficiency. This sudden withdrawal from their beautiful island was a devastating blow and a great anti-climax after two centuries of hard endeavour and discipline.

During the Rhodian years, this multi-national «elite» body of men had successfully combined the military traditions and codes of chivalry of Europe to the highest ideals of their essential monasticism. They were soldiers, sailors, rich men, «poor» men and now rulers; but still monks, dedicated to Christ, the sick and the poor. Free from the shackles of the political intrigues and parochial squabbles which characterized Medieval Europe, the Knights Hospitallers were «sovereign» in their territorial home. They enjoyed a measure of independence (often accompanied by displays of arrogance) which was much envied in some quarters of Europe itself. Born of the Crusades, they had outlived and eclipsed the other military orders like the Knights Templars and the Teutonic Knights. They had successfully adapted to change by diversifying their roles. The Knights Hospitallers were as versatile as they were unique in Christendom.

Yet, the Order was curiously very much a microcosm of the same Europe of the Middle Ages. Its composition, its structure, codes of conduct and even its creative energy were a reflection of the excellence and the excesses of Renaissance Europe. Its strength was drawn from the same power bases which sustained the Christian Kingdoms of that great continent; those of the noble barons, unified only when in conflict with Islam. The Order of St John symbolized that worthy cause. The Knights were the dependable champions of Christendom.

No wonder then, that as the Order's fame and fortunes rose, there was never a shortage of recruits from among the nobility of Europe. Such was the prestige of the Order that boys' names were put down at birth for entry as Knights of Justice. French recruits had to show evidence of four generations of unblemished noble descent; the Germans, eight generations.

After this aristocratic «elite» came the Conventual Chaplains who, although not necessarily of noble lineage, still had to be of «respectable» origins. They served in the churches, the hospitals and in the galleys. Lastly came the Servants-at-Arms, recruited mainly from the local populations to act as serving brothers, clerks, nurses and as soldiers and sailors.

The Order of St John was divided into eight groups or «Langues» formed of Knights of like nationality. At the time of the fall of Rhodes, the Langues were those of Provence, Auvergne, France, Aragon, Castille, Italy, England and Germany. Each had its own house or headquarters called the «Auberge» which provided communal facilities for the fraternity.

The wealth of the Order, accumulated over the years was now legendary. It owned land and estates spread throughout Europe. Grouped into Priories and Commanderies, they were all administered by the Order. A novice entering as a Knight of Justice would bring in with him a « dowry ». On his death, his estate would pass in its entirety to the Order. This system for the accumulation of wealth was not only self-perpetuating. It was also inflation proof. It is said that at one stage the Order was as rich as the Church. At the time of Rhodes, as many as 656 commanderies in Europe were under its control.

Heading the Order was the Grand Master. His authority was absolute and supreme. Elected for life by the Knights themselves, he dominated the affairs of the Order. His allegiance was to the Head of the Church, the Pope himself. The Grand Master ruled through the Supreme Council, composed of the heads of the Langues (called Piliers), the Bishop, and the Conventual Bailiffs who were honorary Knights residing in the Convent.

Specific duties and areas of responsibility were assigned to the Piliers of each Langue. Thus the Pilier of France was the Grand Hospitaller. That of Auvergne, known as the Grand Marshall was responsible for all military affairs. The Grand Admiral was usually an Italian. The Pilier of Provence was the treasurer or Grand Preceptor. The Aragonese were the «quartermasters» of the Order. The Langue of England was in charge of the cavalry and the coastal defences. Its head was known as the Turcopilier. The Germans were responsible for the fortifications. Lastly, dealing with all legal matters and Acts of the Order was the Grand Chancellor, from the Langue of Castille. In times of seige, each Langue was also assigned defined areas within the fortifications to defend.

The system worked well and while a certain amount of squabbling would take place among the various Langues, it also generated a healthy competitive spirit and rivalry among the Knights, which was good for efficiency. The autocratic rule of the Grand Master would always ensure that a high level of discipline was maintained. As each successive Grand Master displayed equally high standards of leadership, the Order progressed and prospered. Its role and destiny were never questioned. Nor were they ever in doubt.

Such was the state of the Order of St John when the Rhodian debacle occured. However honourable the terms of their evacuation from Rhodes may have been, the expulsion of the Knights was a blow to their pride and prestige. As Grand Master Philippe Villiers de L'Isle Adam and his Knights sailed out of Rhodes for the last time, their spirits may have been somewhat dampened. But their resolve was firm. The Order had to be saved. A new home would be found from where they could continue their struggle against Islam. The militancy of the Knights was still their strength.

For seven years they were to wander without a permanent home; first to Civitavecchia, then to Viterbo and lastly to Nice. Away from the safe remoteness of the east, the Order now fell easy prey to the politicking and power rivalries between Philip I of France and Charles V, Holy Roman Emperor and King of Spain and Sicily. In vain did L'Isle Adam try to persuade them to help him retake Rhodes. When Charles V suggested Malta as an alternative, the question of sovereignty and non-alignment of the Order to either of the Christian powers proved to be a stumbling block at first. It was Pope Clement VIII who finally saved the day. By 1529, the dust had settled over his dispute with Charles V and he was now in a better position to bring matters to a head. With nowhere else to go, L'Isle Adam was compelled to accept Malta.

In October 1530, the Knights entered the Grand Harbour and dropped anchor beneath the fortifications of Castello à Mare and Birgu.

The Sovereign Military and Hospitaller Order of St John of Jerusalem, of Rhodes and now of Malta, had come home. Despite its reluctance to settle here, it would be here that the Order would win its greatest laurels over Islam and reach the zenith of its achievement.

1. Bradford: The Shield and the Sword. Op. Cit pp 109-110
2. Bradford: Ibid. p 90

THE GOLDEN ERA OF THE KNIGHTS OF MALTA

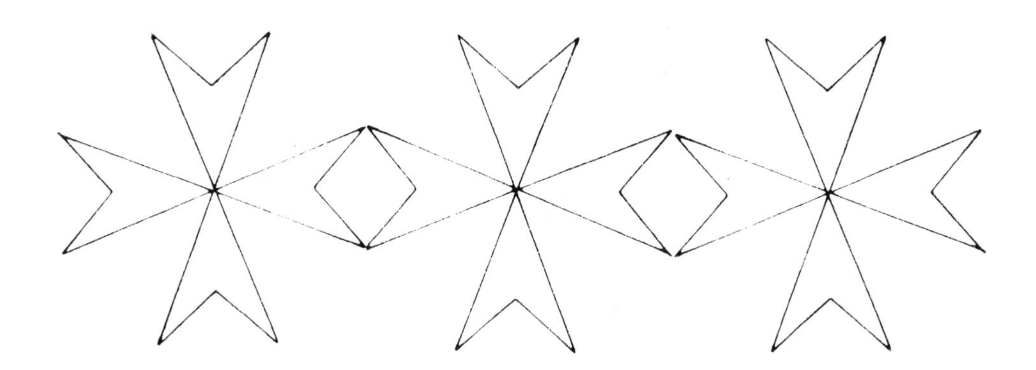

PRELUDE

At first, the Order of St John was most unwelcome in Malta. The Maltese immediately resented the arrival of the Knights. For a start, they were hardly consulted before the Islands' were enfoeffed to the Order by Charles V. They regarded this new development as a violation of the promise given to them by King Alfonso and contained in the Royal Charter of 1428 to the effect that Malta would never be seperated from the Royal demesne. Fearing the loss of their rights and privileges, the Maltese nobles in particular were incensed with this sudden «take over » of the Islands by the Knights whom they condidered to be arrogant and aloof. Indeed, their fears that the authority of the Popular Councils or Universitates would be undermined, and eventually eroded, were to be proved right.

Still harbouring hopes of a possible return to their beautiful Rhodes, the Knights Hospitallers were equally not enthusiastic about Malta at first. They had reluctantly found themselves in an Island which was poorly fortified, under-populated, lacking natural resources and industry and which was heavily dependent on large scale imports from Sicily. Its only saving grace was the Grand Harbour and the other natural harbours in its perimeter. If properly fortified, they could be excellent assets. It was therefore here at Birgu that they based their headquarters; much to the relief of the jealous nobility residing at Mdina. Although determined not to have any truck with the Order, their hostility was thus considerably placated. The population of Birgu was traditionally more cosmopolitan, industrious and eager to work with the new masters.

Soon, the Maltese were joining the Knights at sea in their incessant warfare against the Turks. It is said that in 1535 they took part in the expedition mounted by Charles V against Heyradin in North Africa when 10,000 Christian slaves were liberated. They were also heavily involved in the other campaign against the city of Algiers, then a hot-bed of piracy, which ended abruptly in disaster when the Christian fleet, of which they formed part, sailed into a violent storm.

All the African coast from Morocco to Egypt, along with Syria, Mesopotamia, Kurdistan, Armenia, Hungary, Rhodes and later Cyprus were now in Turkish hands. The Ottomans were the dread of the Christian powers bordering the Mediterranean. The Barbary corsairs, who were the Turkish vassals of the region, continually menaced the sea lanes of the central Mediterranean, raiding, looting and enslaving prisoners along the coasts of Spain, Italy, Sicily and Malta. Although under attack, the Maltese often came to the rescue of their neighbours against the threatening onslaughts of Islam. Neither side ever dreamt of peace. Malta was now becoming of supreme strategic importance for the control of the Mediterranean against the alarming growth of Muslim power.

In 1547 the Turks made an unexpected attack on Malta and Gozo taking many prisoners. However, their invasion in 1551 was more serious. Landing at Marsamxett, they advanced towards Birgu but, taking careful note of the fortifications of the harbour areas, they hesitated and decided to attack Mdina. On their way they ravaged the countryside and villages. Mdina also appeared to be too well fortified to warrant a successful attack. The Turks thus moved on to Gozo which was clearly defenceless. It is said that by the time the Turks had finished with it and withdrew, some 5000 Gozitans had been captured and enslaved.

This set-back finally convinced the Order that more had to be done to defend the Islands. The fall of Tripoli, also in 1551, put paid to any hopes of a return to Rhodes. It was now clear that the central Mediterranean had to be defended. From now on the Order was compelled to put all its resources and energy at the disposal of this region. Plans were made for the strengthening of the fortifications in the Grand Harbour areas of Malta. It was suggested that a new fortress city should be built on Mount Sceberras, the cliffed peninsula which commanded the entire harbour areas around its sides. This would of course take time, which regrettably was not on Malta's side. The pressing urgency lay in resisting another strong invasion by the Turks. A star shaped fort was therefore built at the tip of Sceberras to guard the main harbour entrances. It was named after St Elmo, patron saint of seamen. The old Castello à Mare, now renamed Fort St Angelo,was strengthened along with the adjoining town of Birgu. On the adjacent peninsula, Fort St Michael was erected, serving also to defend the new town of Senglea.

The economy of the Islands started to revive with all this feverish activity. Agricultural produce was on the increase,although Malta still relied heavily on the importation of grain from Sicily. The trade gap was largely bridged by the vast sums of money which poured into Malta from the Order's Priories in Europe. The Knights did not only have to maintain the hospital at Birgu, their Convent and the fleet of eight war galleys; they also had to finance the building of the new fortifications and the purchase of adequate war supplies and munitions. Sadly, the revenue of the Order had been severely cut back when, with the Reformation, some German Priories had been lost. Henry VIII of England had also suppressed the English Langue. There was also the continuing rivalry between Charles V and Philip I of France which the Order had to contend with. When France entered into an alliance with the Turks in order to counter-balance the excessive power of the Holy Roman Emperor, the Sicilian Viceroy became distrustful of the French Knights who largely dominated the Order. He therefore sought to exercise some control over the Order by keeping Malta short of food supplies. This led to famine and much misery and hardship among the Maltese.

A sudden ray of hope and inspiration appeared on the Maltese scene when a Frenchman, Jean Parisot de la Valette was elected Grand Master of the Order. With him, and through him, the golden era of Malta commenced. It would be best to let a contemporary of his, Francisco Balbi da Correggio, a wandering poet and soldier who served throughout the Siege of Malta, describe the greatest Grand Master in the annals of the Order. In 1565, he wrote:

«After the death of Claude de la Sangle the choice of a Grand Master fell on Jean de la Valette, who now rules the Order. He is a French Knight of the Langue of Provence, a good Christian, as brave as he is wise, and a good leader. Before he attained to the highest dignity of his Order he held many inferior charges from commander of a single galley to the command of all the galleys. He had been Bailiff of Lango, governor of Tripoli, and had taken part in the Siege of Rhodes.

He was elected Grand Master in the month of August of the year 1557, and, so that all may know his worth, I shall briefly describe his qualities for the benefit of those who never met him.

In 1557 his age was 67; he is tall and well made, of commanding presence, and he carries well his dignity of Grand Master. His disposition is rather sad, but, for his age, he is very robust, as we have seen in the past trouble. He is very devout, has a good memory, wisdom, intelligence, and he has gained much experience from his career on land and sea. He is moderate and patient and knows many languages. Above all he loves justice, and he enjoys the favour of all the Christian Princes. It required a man of his courage and wisdom to resist the most cruel attack of Soliman, this year, as you will now read». [1]

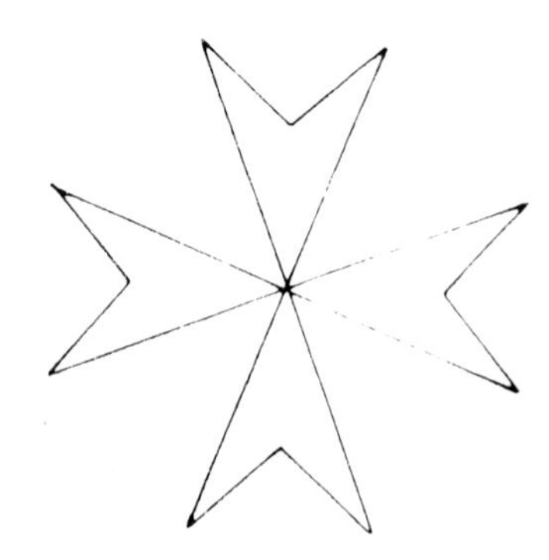

1. Francesco Balbi di Correggio: The Siege of Malta 1565 (Published by O.F. Gollcher and O. Rostock Copenhagen 1961) p29

THE GREAT SIEGE OF MALTA

In the autumn of 1564, Suleiman the Magnificent, now an old man, was bitterly regretting the clemency he had shown towards the Knights of St John forty two years earlier, when as a young Sultan, he had spared the Order on his conquest of Rhodes. He had come a long way since then, but the Knights had remained his implacable enemies. He was now determined to destroy them once and for all. He would call in his fleets and his crack military units and assemble all the resources of his vast empire and pit them against Malta, the last refuge of those «sons of Satan» who «shall for their continual piracy and insolence be forever crushed and destroyed».[1]

News of the impending attack on Malta by the Ottomans reached La Valette quickly enough. He too realised that this would be the Order's last stand against their mortal enemy; a fight to the finish. Accordingly, he made immediate preparations. He recalled all the Knights from Europe. He then embarked on a systematic plan to strengthen further the fortifications concentrating on the Grand Harbour. Next, he stockpiled the Island with the latest military hardware of the day and imported adequate food supplies from Sicily to sustain the population for a long siege. Regrettably, his pleas to the Christian Princes for manpower and material support largely fell on deaf ears. Only the Viceroy of Sicily sent in some 1000 Spanish infantry. Because La Valette only had about 540 Knights and Servants-at Arms, it was clear that the burden of resistance would have to be borne by the Maltese population, of which about 4000 men had been formed into a local militia.

On the 18th May, 1565 the Turkish fleet appeared off Malta. It numbered some 200 vessels, armed to capacity with about 40,000 troops. In charge of the Army was Mustapha Pasha, a veteran of Rhodes and a hero of the Hungarian campaign. Suleiman's own son-in-law, Piali Pasha was the Admiral of the Fleet. Under their commands, they had the finest and most seasoned troops drawn from the length and breath of the Ottoman Empire.

Sailing unopposed into the refuge of the large southern harbour of Marsaxlokk, the Turks lost no time in landing their troops. Soon they were encamped at Marsa, the extreme land end of the Grand Harbour. Piali was clearly unhappy to site his precious fleet at Marsaxlokk. He prefered the protection of the smaller harbour of Marsamxett. Regrettably, this was within range of the guns on Fort St Elmo, which had first to be neutralised before he could redeploy his fleet there. At his insistence, the Turkish command decided to make Fort St Elmo their first target. Siting their artillery on the heights of Mount Sceberras, they commenced on the systematic bombardment of the little fort. The walls were breached often enough but the gallant defenders could not be removed. By night La Valette would send in re-enforcements from Birgu across the harbour and ferry the wounded safely back. However, it was clear that the fort was doomed.

The greatest Barbary corsair of them all now appeared on the scene. Torghoud Rais, better known as Dragut, who had led the Turkish invasion of Malta and sacked Gozo in 1551, conqueror of Tripoli, the scourge of the Mediterranean and most trusted field commander of Suleiman, arrived in Malta and immediately took effective command. He disagreed with the strategy of Mustapha and Piali. The Turks should have first secured Mdina and then concentrated on the main fortifications in the Grand Harbour; Forts St Angelo and St Michael. The little Fort at St Elmo was the least important. Committed as they now were to it, he re-sited the Turkish batteries for a more effective shelling of the fort. He also sent out night patrols in the Grand Harbour thus severing the defenders lifeline with Birgu. On June 22nd, after 31 days of constant siege, St Elmo fell. As the Turks rushed through, Dragut was killed by a ricocheting shell from St Angelo.

The Turks had paid a high price in terms of men and morale to overwhelm the tiny fort and thus gain access into the dubious safety of Marsamxett harbour for their flotillas.

Mustapha now turned his guns towards Birgu and Senglea. In a vain attempt to undermine the morale of the Order, he nailed the de-capitated bodies of some captured Knights on to makeshift wooden crosses and floated them downstream across the harbour waters in the direction of Fort St Angelo. La Valette's immediate response was ruthless ; he fired the heads of Turkish prisoners through his cannon across to the enemy's positions. It was clear that the subsequent stages of this battle were to be conducted with unparalleled ferocity and savagery.

The Janissaries.

Mustapha blockaded the two peninsulas of Senglea and Birgu from their landward ends and from the sea. Across land, he dragged some of his galleys from Marsamxett to the Marsa end of the Grand Harbour ; being unable to sail directly through the harbour and risk being fired at by the guns of St Angelo. From Marsa, he could now launch a combined land and sea attack on Senglea after first subjecting the region to a concerted bombardment from all sides. However, the Maltese had erected underwater obstacles all along the shallower reaches of the peninsula in anticipation of such an attack from the sea. As the enemy boats were trapped in their advance, Maltese swimmers now engaged the Turks in a hand to hand battle in the water. The Turks were massacred.

When the Janissaries, the cream of the Turkish troops, tried to gain entry from the lower reaches of the other side of Senglea, by way of a diversion to the main fighting zones, they were surprised to come under fire at point blank range from a concealed battery on the waters' edge of Fort St Angelo. They were annihilated. Things were going badly for the Turks as the siege moved inexorably into its third month and into the sweltering heat of the Maltese mid-summer. The Turks continued to bring all their cannon to bear on the fortifications. At one point Senglea was on the point of being taken when surprisingly a retreat was sounded. The Turkish base camp at Marsa had been attacked by the Order's cavalry from Mdina to cause a diversion. Mistakenly, the Turks thought that a relief force had arrived on the Island. The momentum of the siege soon got under way again in earnest. Now Birgu came under great pressure. When an exploding mine created a gap in the main bastion, the Turks surged forward. The Crescent was seen flying on its ramparts. Rallying his men, 75 year-old La Valette quickly led a counter-charge in person and was wounded in the leg. His leadership and example further inspired the defenders.

The long drawn out battle was having a telling effect on both sides. The Knights were reeling under the considerable fire-power hurled at them. However, the Turks had suffered appalling losses and their morale was now sagging.

On 6th September, a relief force from Sicily did actually arrive and landed in the comparative safety of Mellieha bay. It numbered only about 8000 soldiers. However, Turkish estimates of its strength were much exagerated. Surprisingly, they called a general retreat. Quickly, they withdrew from the blood spattered walls of the great bastions they were on the point of taking. Their evacuation of the Island was swift. On the 8th of September, the church bells of Birgu and Senglea were pealing the sweet sounds of victory. The siege of Malta was over.

Some 7000 Maltese and Spanish and 250 Knights had perished. However, the Turkish losses estimated at around 30,000 were far greater.

The significance of the great victory became all too clear in Europe. The Ottoman Empire had received a body-blow from which it was never to recover. Its advance into Europe had been checked. Suleiman the Magnificent had been humbled. Christendom had been saved. Malta was now acclaimed for her heroism. With this glorious victory, she had now featured on the international stage. Voltaire would later comment - « Nothing is better known than the siege of Malta ». Malta had also entered the realms of modern history.

«Malta of gold, Malta of silver, Malta of precious metal,
We shall never take you!
No, not even if you were as soft as a gourd,
Not even if you were only protected by an onion skin!
And from her ramparts a voice replied:
I am she who has decimated the galleys of the Turks -
And all the warriors of Constantinople and Galata!» [2]

From a sixteenth century Cypriot ballad.

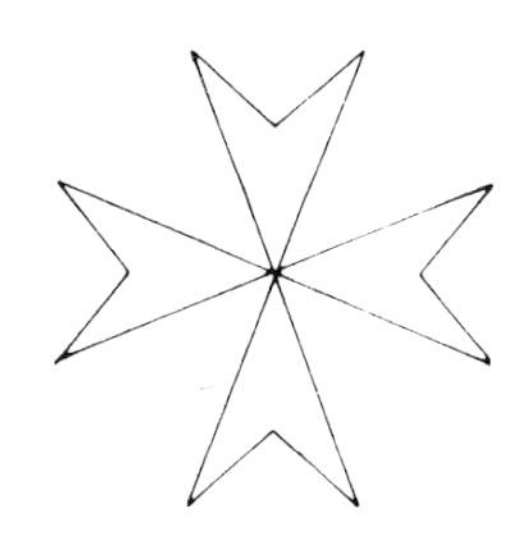

1. Bradford: The Shield and the Sword. Op. Cit. p. 143
2. Bradford: Ibid. p. 172

VALLETTA - THE CITY ON THE HILL

As Malta and the Order of St. John gloried in the famous victory over the Ottomans, it was immediately clear that, true to its tradition, Islam would soon seek its revenge. The Island would be in no position to resist another serious attack unless something was done urgently to restore and improve the ruined fortifications. Because the battered Island was now clearly defenceless and utterly vulnerable, some Knights even wanted to evacuate it.

Grand Master Jean de la Valette, the hero of the Siege was adamant that the Order should remain in Malta. It would be shameful for the Knights to abandon the Islands just when they were being acclaimed throughout Europe as the shield of Christianity. He therefore immediately resurrected the project for building a fortified city on Mount Sceberras. By early January 1566, outline plans were drawn up by Francesco Laparelli de Cortona, an engineer and an expert in military fortifications who had been sent to Malta for this purpose by the Pope. Soon he convinced the vacillating Knights that their future would be better served and more secure in the great citadel which he proposed to erect on Sceberras. On the 28th March 1566 the foundation stone of the new city was laid. Its inscription in Latin reads as follows:

> *«Fra Jean De La Valetta, Grand Master of the Hospitaller Order of Jerusalem, mindful of the danger of which, a year before, his Knights and the Maltese people were exposed during the siege by the Turks, having consulted the heads of the Order about the construction of a new city and the fortifying of the same by walls, ramparts, and towers sufficient to resist any attack or to repel or, at least to withstand the Turkish enemy, on Thursday the 28th March 1566, after the invocation of the Almighty God, of the Virgin Mother, of the Patron Saint John The Baptist, and of the other Saints, to grant that the work commenced should lead to the prosperity and the happiness of the whole Christian community, and to the advantage of the Order, laid the foundation stone of the city on the hill called Sceberras by the natives, and having granted for its arms a golden lion on a red shield wished it to be called by his name, Valletta».* [1]

At once it was clear that the cost of the completed project would be appalling. Immediately, a call went out to the Christian Princes for financial support. Large donations, particularly from France, were received from a Europe willing to express its gratitude for the heroic stand of the little Island in the Mediterranean. Pope Pius V even proclaimed a jubilee Indulgence to benefit those faithful who would contribute towards the building of the new bulwark of Christianity.

Valletta, with its formidable walls and bastions along the sheer cliffs of the entire peninsula .would dominate the harbour areas on both its sides. Its very impregnability on high ground would in itself act as a «deterrent» to the enemy. Never again would the three cities of Birgu, Senglea and Bormla be exposed to the merciless cannonade of enemy guns on Sceberras. The defensive perimeters of the harbour areas would all be protected by walled extensions and fortifications. St Elmo, guarding the entrances to the harbours would be enlarged. The naval stronghold within the Grand Harbour would be protected by Forts St Angelo and St Michael. The Three Cities would be contained by an impressive network of defences to shield them from the landward ends. In time, the defensive scheme would include the new fortresses at Ricasoli opposite Fort St Elmo, and Fort Tigné guarding the entry into Marsamxett harbour. Indeed, within that harbour, the little islet, later to be known as Manoel Island, would itself contain a fort.

If Laparelli merits the credit for not only winning approval to commence work on the new city but also for its design and layout, then the supreme accolade for its building must go to his Maltese assistant, Gerolamo Cassar. With the death of La Vallette in 1568 and the departure of Laparelli from the Island, it would be Cassar who would give Valletta its sombre elegance and character. He received much support from the new Grand Master, Pietro del Monte who was equally as enthusiastic about the project. It was Cassar who designed the great buildings, St John's Co-Cathedral, the Parish Church of St Paul, that of St Augustine besides other great houses. He even designed the «bakeries and the windmills of Valletta» [2] which regrettably have now disappeared. Cassar contributed much towards the high traditions of Maltese architecture which persist to this day.

By 1571, Del Monte had formally moved the Order's Convent to the new capital, even though it was still in the course of construction. That year also saw the destruction of the Turkish Fleet in the Battle of Lepanto which clearly brought some relief to the Islands from the hung threat of Turkish attack. However, the building of the fortifications continued in earnest with each successive Grand Master. By 1582, Valletta was clearly taking shape and it was attracting more residents. Strict regulations had been laid down to ensure that its buildings would be of a high standard. Indeed, only palaces were allowed in certain streets. Even street corner sites had to be suitably decorated.

The expansion of Valletta was rapid once the Order moved in. Its social fabric was gradually formed around the main streets; Strada San Giorgio (later Kingsway and now Republic Street), Strada Merchanti or Merchants Street and the more residential Strada Forni, now Old Bakery Street. There was to be no closed «collacchio» in Valletta, as there had been in Rhodes, to isolate the dwellings of the Knights from the other houses of the city. Many Knights now lived in their own private houses and only dined in their Auberge a few times during the week. Not far from the now enlarged Fort St Elmo, was built the great hospital of the Order in 1574. Then known as the «Sacra Infermeria» it became world famous for the quality of its medical care.

Valletta became the impregnable fortress of its design. It stood in contrast to the ancient citadel of Mdina, then re-named Citta Notabile. Although Mdina still remained the seat of the Università, the new capital of Valletta soon eclipsed it. The Università was gradually reduced to a powerless municipality. Soon Mdina became known as the old city - Citta Vecchia. Considered indefensible, its importance declined. However, many fine buildings were erected during the time of the Knights. They are largely baroque in style, like the Magisterial Palace by Barbara, and the Cathedral with its fine dome by Lorenzo Gafa. The population of Mdina dwindled. Its only inhabitants were ecclesiastics and the Maltese noble families who, in their ancestral homes, never collaborated much with the Order and kept away from the hustle and bustle of emergent Valletta. The serenity of Mdina, the silent city, is preserved to this day.

In time, Valletta would become a centre of art and culture. In an ostentatious age, the Knights were as prone as anyone else to embellish their great palaces and their churches. They were able to acquire the services of the best painters and sculptors of Europe. In 1608, Caravaggio came to Malta. His great masterpiece - the Beheading of St John - now graces the Oratory Chapel of the Conventual Church in Valletta. At a later stage, Mattia Preti, working in oils directly on the Malta stone painted the vaulted ceilings of that Church. He also designed the gilt carvings on the arches and walls. The floor was covered with mosaic marble slabs placed over the tombs of the more distinguished Knights. The Maltese sculptor Melchiorre Gafà and famous Italian sculptors all contributed towards the magnificence of «St John's». Grand Master Perellos was also to commission the famous Flemish Tapestries woven in Brussels from cartoons by Peter Paul Rubens, which can now be seen in the museum of the Co-Cathedral.

The Palace of the Grand Master became a veritable treasure chest of art collections with the most fashionable furniture of the period and rich decorations. To it, Perellos also donated the Gobelins Tapestries of the series «Les Teintures des Indes» still to be seen in the Council Chamber. In the Hall of St Michael and St George, scenes of the Great Siege by Matteo Perez D'Alleccio decorate the walls.

The quiet facades of Gerolamo Cassar's late 16th Century buildings gave way to the more ornate Baroque style which Bernini had introduced in Rome. In Malta, it pervaded everywhere in the late 17th Century - to the point that Valletta could now be accused of being essentially a Baroque city. However, Maltese Baroque was perhaps more restrained in its use when compared to its equivalent in Spain or Italy. During these productive and prosperous times, the names of Maltese architects like Barbara, Cachia, Zerafa, Bonnici and Dingli emerge as giants on the local scene for the excellent examples of their work in Valletta and elsewhere. Domenico Cachia is celebrated for his new Auberge de Castille which replaced the old one by Cassar. Today, it is used as the Office of the Prime Minister. Cachia also built the Parish Church of St Helen in Birkirkara - perhaps the finest example of Baroque in Malta. Giovanni Barbara became famous for the Church of St James in Merchants Street, the most «Baroque» street in Valletta.

Valletta became a hive of creative activity. Painters, sculptors, cabinet makers, gold and silversmiths, tailors, shoemakers and other craftsmen all contributed towards its grandeur and charm. Abetted by the Knights, the Maltese revelled in this «rennaissance» of art in Malta. On the literary scene too, Maltese scholars and writers made their mark. Gian Francesco Abela published his «Della Descrizione di Malta» in 1647. Nine years earlier, Salvatore Imbroll had published in Naples another fine work; the «Specula Melitensi Encyclica». Other notable writers were the Maltese Grammarian, Agius de Soldanis; the Philosopher and Theologian Domenico Bencini and the Biblographer Nasio Mifsud. In the field of music, the names of Michelangelo Vella, Francesco Azzopardi and later, Nicolo Isouard, became famous. Finally the works and idealism of the great patriot, Mikiel Anton Vassalli, perhaps the Island's first Democrat in the modern sense, cannot be overlooked.

«Give me time and I will give you life» [3] wrote Laparelli in the early furious years of the building of Valletta. Together with Gerolamo Cassar, he had fulfilled his promise. The city that had arisen out of the ashes of war on the rubble of Mount Sceberras was to become the heart of Christian Malta, a microcosm of the excellence and magnificence of Occidental ascendant Europe, an impregnable fortress city overlooking the best natural harbours in the central Mediterranean, the proud citadel of the Sovereign Military and Hospitaller Order of St John and the famous capital of the Maltese Islands.

Forever revered by the Maltese, Jean Parisot de la Valette, lies buried in the city that bears his name. His tomb bears the following inscription:

> *«Here lies La Valette, worthy of eternal honour. Once the scourge of Africa and Asia from whence he expelled the barbarians by his Holy Arms, he is the first to be buried in this, his beloved city which he founded».* [4]

1. Quentin Hughes: Fortress Architecture and Military History in Malta (Lund Humphries, London. 1969) p. 61
2. Hughes: Ibid p. 79
3. Hughes: Ibid p. 73
4. This epitaph translated from the Latin was composed by La Valette's personal secretary and Knight of the Langue of England, Sir Oliver Starkey, himself now buried next to La Valette in the crypt of St Johns Co-Cathedral in Valletta. He is the only Knight, other than a Grand Master to be buried in the crypt.

THE YEARS OF CONSOLIDATION

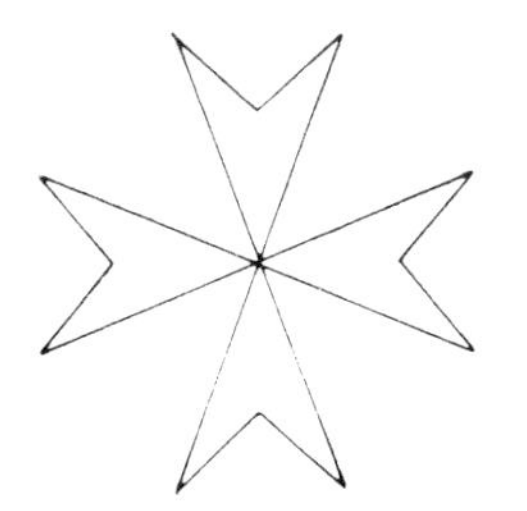

At the entrance to the Gulf of Corinth, a great sea battle took place in 1571. Here, the combined European navies of Spain, Venice and Genoa, along with three galleys of the Order of St John, scored a resounding victory over the Ottoman fleet. In the Battle of Lepanto, as it came to be called, the Turks suffered their first serious defeat since their failure to take Malta six years earlier. While this battle contributed significantly towards the ultimate decline of Turkish power in the Mediterranean, it did not immediately remove their presence from the region. As usual, they came back with a vengeance. With a re-organized fleet, they soon regained control of the Mediterranean. In 1574, they recaptured Tunis and with the help of the Berber pirates, they were soon in possession of Morocco, Algeria, Tunisia and Tripolitania. As the Mediterranean became infested with corsairs from the Barbary coast, there was total anarchy at sea. The Turkish menace to Christian shipping grew stronger and stronger. Their sporadic raids on the coasts of Malta continued unabated. The Knights of St John were thus compelled to maintain their vigilance over the Islands for many years to come. While Malta was now in a better position to withstand an attack on her, the Order relentlessly pursued its programme to provide better security for the Islands. For the next two centuries, work on new fortifications, extensions to lines of defence and on the erection of many coastal defence towers would continue until a comprehensive deterent effect would be achieved.

At sea, the galleys of the Order took to the offensive. Each season, they would sail out of Malta and generally patrol the region. At first, the Knights sought to protect and come to the rescue of Christian shipping. Soon, the plundering of that of the enemy would become more commonplace. The rewards were too great to resist. Maltese pirates, often flying the Order's flag, flourished in the wake of the Order's lead. It is estimated that half of the Islands' male population was dependent on piracy in the 17th Century.[1] Maltese corsairs made spectacular gains over the length and breath of the Mediterranean. Sometimes they provoked many protests and retaliatory threats from their victims which often included Christians. A serious incident arose in 1669 when some Maltese pirates attacked the British Consulate in Cyprus.[2] This caused much embarrassment all round, particularly to the Order which had always allowed English shipping free passage into Malta. In time, restrictions had to be imposed on the Maltese corsairs to keep them under control.

As the Order of St John settled and consolidated its position in Malta, it was clear from the start that the religious fervour which had so inspired the Knights in Jerusalem and in Rhodes would slowly evaporate in their new home. The seeds of moral decay and decline within the monastic fabric of the Order were really sown with the arrival of a new breed of Knights to replace those slain during the Great Siege. The Europe which they had come from was different to that of the heroic and crusading days of old. The great humanism normally associated with the Renaissance was fast disappearing. The excitement and sense of experiment and adventure which had led to the discovery of the New World was losing its vigour. Europe was sadly in a state of disorder. The counter-reformation, designed to combat the rift between the Catholic south and the Protestant north, had paralysed the spirit of the Renaissance. With France in the throes of civil war and Italy plagued by banditry and general anarchy, the political scene was largely dominated by the two «super-powers» - Catholic Spain, busily expanding its colonial empire and Elizabethan England, slowly emerging as a great maritime power but equally busy «creating» Catholic martyrs. Everywhere, the high morality and fervour of a once militant Christianity were being replaced by a more relaxed and less rigid adherance to the old dogmas. The affluent young in Europe were restless, rebellious and disinclined to perpetuate the hollow codes of chivalry now thrust upon them. No wonder then, that the new recruits into the highly respected Order of St John were finding it difficult to reconcile the all pervasive licentiousness of the European courts with the chastity, poverty and obedience expected of them by virtue of their vows with the Order.

The presence of many prostitutes in Malta, largely destitute wives or widows of «absent» husbands on piratical forays, further aggravated the situation in Malta as each Grand Master sought to maintain standards and to root out the near total disregard of celibacy among the ranks of the Knights. It was Grand Master Jean L'Evêque de la Cassière (1572-1581) who first made a determined attempt to curb the many excesses of the Knights' behaviour. Also unhappy with some cases of heresy in the Order, and with other cases of corruption within the Church in Malta, he called for a Papal Inquisitor to sort matters out. This only created more confusion. The Knights rebelled and, declaring La Cassière incompetent, went on to imprison him in Fort St Angelo. They now proposed to install a new Grand Master - Martin Lescout, known as Romegas. The Pope would have none of this. Summoning both men to Rome, he roundly chastized them. But in the end, La Cassière was vindicated and re-instated as Grand Master. However, he was never to rule again in Malta. By a twist of irony, both he and Romegas died in Rome. La Cassière was brought to Malta and was buried in the Conventual Church of St John. It was a fitting gesture, since it was through his generosity that it was built and its completion marks his greatest achievement in Malta.

The field was now clear for another controversial Grand Master to emerge - the Knight from Provence, Hughes Loubenx de Verdalle. His mastership lasted for 14 years during which the Maltese witnessed the slow erosion of the rights and privileges of their popular councils and Università under the heavy hand of his despotic administration. Dynamic, iron willed, even impressive, Verdalle was also intensely vain and wordly. He surrounded himself with all the trappings of power and pomp with which he associated his high office. The complete autocrat, he was always in command. Under him there would be no rebellions; not even from the Spanish and Portuguese Knights who despised him. No doubt, his vanity reached its apogee when Pope Sixtus V made him a Cardinal.[3] As a Prince of the Church, his pretensions to grandeur now increased. Overlooking the only wooded area in Malta known as Buskett, he built for himself a luxurious country retreat for which he became famous. Verdala Castle, set in the midst of pines, citrus trees and aromatic shrubs, became his lordly manor. Here, he would dine to the sound of sweet music and entertain his friends and dignitaries in a superlative manner. This would all be happening when Malta was in the throes of a severe famine, soon to be followed by the plague which claimed the lives of 800 Maltese in 1592.[4]

However it has to be said that Malta and the Order benefitted from the riches that Verdalle amassed during his lifetime. On his death in 1595, he bequeathed his entire estate (including that element which he was entitled to dispose of elsewhere) to the Order. Verdala Castle today retains its pomp and dignity. It is now used by the Government of Malta to accomodate visiting Heads of State and other dignitaries.

More benevolent and compassionate was the rule of the succeeding Aragonese Grand Master Martin Garzes (1595-1601). During his brief period of office, he made several attempts generally to improve the lot of the Maltese and to protect the poor. At that time, the Jews in Malta were enjoying a near-monopoly of the credit business in the Islands. Garzes immediately curbed the abuses in connection with the high interest rates which they were charging. He instituted the first official «pawn shop» in Malta, known as Monte di Pietà. He also showed a particular interest in the Island of Gozo. To give better protection to its inhabitants, Garzes modified the fortifications of Rabat and built the Gran Castello. Overlooking the Comino channel at Mgarr, he also erected a small fort.

With the advent of the 17th Century, Malta and the Order of St John moved inexorably into a period of unprecedented progress and prosperity. Through a succession of able and enlightened Grand Masters, the Maltese Islands were to acquire a reputation for orderliness and good government which were in sharp contrast to the anarchy and depression which prevailed in Southern Italy or in the order islands of the Mediterranean. Malta was now flourishing. Valletta and the Three Cities of Birgu, Senglea and Bormla were booming. At Birgu, the Order's small shipbuilding yard, mainly used for the repair and fitting out of the galleys, was expanded. The successful maritime exploits of the navy, combined with the additional booty from the forays at sea of the Maltese corsairs, were bringing in much wealth to the Islands. As a result, Malta became an important slave market. Many slaves were captured at sea during the «Corso» - the thrice yearly cruises of the galley squadrons. They were used as oarsmen in the galleys, as «cheap labour» on the building sites and as servants in the great houses. Slaves were also exported to many parts of Europe.

The prosperity and relative affluence of Malta in the 17th Century was immediately reflected in a dramatic growth in the population. From a figure of about 20,000 in the early half of the 16th Century, the population had risen to 50,000 by 1632. Later on, by the time of the Order's departure from Malta, that figure would be doubled. The growth in population was perhaps most rapid during the successive periods of rule of Grand Masters Alof de Wignacourt (1601-1622), Antoine de Paule (1623-1636) and Jean-Paul de Lascaris (1636-1657). The drift of the population towards the main urban centres of Valletta and the Three Cities had to be checked many times by the Order. Mainly for this reason, Grand Master de Paule founded a new town between Marsa and the village of Tarxien, now known as Pawla or more commonly, as Rahal il-Gdid. To encourage its growth, «he granted the inhabitants the right to exemption from the payment of debts for a period of time - a sort of moratorium. Hence, the nickname «Il-Midjunin» (the debtors) which has survived to this day,[5] by which the people of Pawla are known. The hazards to health and hygiene, as a result of gross over-crowding in the congested harbour areas, were always foremost in the minds of the Knights. They did their best to disperse the population by encouraging the growth of new towns and villages. When plague broke out in 1676 during the mastership of Nicolas Cotoner (1663-1680) their fears were proved right. The worst hit region was the urban area of the Grand Harbour, and particularly Birgu.

Over the years, each successive Grand Master would leave his personal imprint upon the Islands. Some would become famous for the benevolence of their rule. Others would go down in history as tyrants. Nearly all of them would achieve a kind of immortality by lending their names to the new fortifications and coastal defence towers which sprang up during their mastership. Their busts and banners, arms and emblems would adorn the facade of many a great palace in Valletta and elsewhere. The endless number of community projects and foundations which they initiated and set up, often funded from their own pockets, would also make them famous for all time. All would be remembered for their unceasing warfare against the Turks and for the prompt support they would give to the Christian powers of the day. The Grand Masters also had one thing in common. They were all autocrats who ruled absolutely. In that sense they remained feudal in their attitude towards their subjects. For this reason their relations with the clergy in Malta, then the more educated section of the population, would always be stormy. In time, resentment was to spread among the rest of the population and their authority was to become suspect. Their relations with the nobility, too, were never happy. They never enjoyed their confidence or co-operation. The nobility never forgave the Order for disallowing Maltese entry into the Order as Knights of Grace. This was considered an insult and tantamount to non-recognition of the Maltese claims to nobility. But in the main, there can be no doubt that the Grand Masters served Malta well. Always progressive, in some cases even dynamic, they did improve conditions in the Islands and they set the foundations for Malta's future role in the 19th Century. Through their shrewd diplomacy, they were able to steer Malta out of her backwardness and give her an international status, quite disproportionate to her size.

Malta's smallness always presented a problem. Lacking natural resources and insufficient in food production, she remained heavily dependent on Sicily for her grain. However, great strides were made in agriculture. The Order actively encouraged Maltese farmers to make better use of the available arable land. New crops were introduced. Better irrigation contributed to increased yields. Gozo, always the more fertile island, remained the main source of fruit and vegetables. Because grazing land was scarce, the barley crop, mainly used to feed cattle, became important. Cumin seed, used for medicinal purposes and for spicing, continued to be grown and even exported. Citrus fruits were also exported. Indeed, Maltese blood oranges became a delicacy in Europe. The cultivation of cotton was revived and soon, spun cotton was being exported. The weaving of sail cloth, known as «kotrina» became an important industry.[6] The production of salt was stepped up when new salt pans were laid out at Salini. At one time, attempts were made to grow tobacco, and even the mulberry tree for the production of silk.

The marked feature of the times was perhaps the efflorescence of art and architecture not only in Valletta, but also in the more successful villages which now vied with each other to build and decorate splendid churches in their parishes. The pomp and ceremony, to which the Knights became increasingly addicted, encouraged the trend towards more grandiose buildings and monuments. Their standards were high; and so was the resulting level of artistic acchievement which they sponsored. In the latter part of the 17th Century, this creative effervescence would explode in the burst of the new Baroque movement which so gripped the Islands.

The great initiatives of the Grand Masters are legendary. Early in the 17th Century, Alof de Wignacourt built a great aquaduct to bring much needed fresh water to Valletta from Rabat, nine miles away. This much improved the city's hygiene and enhanced its status. The fountains that he then built at many of its street corners added to the charm of Valletta. Wignacourt also erected many fine fortress towers on the coasts and overlooking the more important bays. The one at St Paul's Bay was named after him. Fort St Lucian at Wied Il-Ghajn and Fort St Thomas at Marsaxlokk were completed during his mastership. Work on similar fortresses and fortifications never stopped. In 1634, Antoine de Paule brought in an engineer to design a new line of fortifications to protect the landfront of Valletta. A century later, when the suburb of Floriana was built outside its walls, it was named after that engineer - Pietro Floriani. Grand Master Jean-Paul de Lascaris (1636-1657) built the Margherita lines to protect the city of Bormla. He also became known for the fine wharfs and warehouses of Valletta. His successor, Martin de Redin (1657-1660) became famous for the great number of castles and fortress towers he erected, at his own expense, during his brief mastership. Amongst these, the better known are Delimara Tower, Ghallis Tower near Salini and the Red Tower near Benghajsa. Grand Master Raphael Cotoner (1660-1663) and his brother Nicolas (1663-1680) who succeeded him, continued works on the fortifications. The latter Grand Master inspired the building of the magnificent line of fortified walls which, extending from French Creek to that of Kalkara, would give landward protection to the Three Cities. Although completed much later, Cottonera Lines, which were largely financed from his purse, today bears his name. The mastership of Nicolas Cotoner was progressive and energetic in many fields. He pioneered the teaching of anatomy and surgery in the Order's hospital. He also set up the «Fondazione Cotoner» which provided funds for the erection of windmills at Bormla, Naxxar, Zebbug and Zurrieq.

If the fortifications of Malta were built to deter and repel any Turkish invasion, the Knights lost no opportunity to fight them elsewhere. De Lascaris is known for his early committment in support of the Venetians when the Turks threatened Crete in 1645. It was a prolonged war and the Order's galleys fought many sea battles on the Christian side. Seven galleys were involved in the victory at the mouth of the Dardanelles in 1656 which resulted in the freeing of about 7000 Christian slaves. One of the heroes of the battle was Grégoire Carafa who became Grand Master in 1680. He remained most militant and during his ten year rule he continually encouraged the Knights to assist their Christian brothers in their conflict with Islam in Eastern Europe. Indeed, the Order played an active part in the successful defence of Vienna in 1683. Under Adrien de Wignacourt who was Grand Master from 1690 to 1697, the Order was still actively fighting the Ottomans in Morea with the allied forces of Austria, Poland and Venice.

As the 17th Century came to a close, it was fitting that the Order would elect the wealthy Raymond Perellos y Roccaful as Grand Master. His mastership well typifies the confidence and standing with which the Order of St John moved during that century. In Malta, he will forever be remembered for his generous donation, out of his own funds, of the priceless sets of Flemish and Gobelin Tapestries to the Church of St John and to the Palace of the Grand Master. Perhaps he best caught the mood of the times in the arched gateway, then complete with drawbridge and ditch, which he built outside Floriana at Porte des Bombes. It bears his carved emblems and an evocative inscription in Latin:

«As I fight the Turks everywhere, here I am secure and powerful».

1. Vella: Op. Cit. Vol. II. p. 327
2. Vella: Ibid. p. 330
3. After the Great Seige, La Valette too had been offered a Cardinal's hat but he declined it.
4. Vella: Ibid. p. 61
5. Cassar-Pullicino: Op Cit. p. 128
6. Cassar-Pullicino: Ibid. p. 139

DECAY AND DECLINE - AND FALL

The confidence and arrogance that underlines the tone of the inscription at Porte des Bombes somewhat betrays the irony that the very safety and security which the Order of St John enjoyed in the Maltese Islands largely rested on the very fact that it was able to inflict defeats on the Turks in the Mediterranean. However, once the Turkish threat started to recede, the «raison d'être» of the Order would be immediately called into question. Its maritime efficiency and success was itself contributing towards its very obsolescence.

Besides, the Mediterranean scene was changing rapidly. There was a revolution in shipping itself. The era of the oared galleys was at an end. Sail was replacing it. Recognizing this development, Grand Master Perellos was very quick to re-equip the Order's navy, at vast expense, with the modern, large round-bottomed vessels and huge canvas sails. Equally well equipped were the powerful navies of the European maritime powers, including those of the newcomers to the Mediterranean - the British and the Dutch, now displaying early signs of arrogance themselves. It was abundantly clear to all that the Ottomans were being out-manœuvered not only at sea but on land too. They were routed in Vienna in 1683 and their grip of eastern Europe was to fade rapidly with the Treaty of Karlowitz of 1699. While these events did not lead to any relaxation of the Order's vigilance over the Mediterranean region, they did prompt it increasingly to take to the offensive with renewed vigour. Each victory would earn them greater laurels. Ironically, each victory would further diminish the Order's role in the Mediterranean.

In Malta the possibility of minor incursions and raids by the Turks and the Berbers always remained. Each Grand Master would guard against this. With each fresh scare - and there were many in the 18th Century - new fortifications would be erected. Because of one such threat during the mastership of the well-loved Antoine Manuel de Vilhena (1722-1736) the relatively unprotected area around Marsamxett harbour was strengthened by the building of a new fort on the islet within it. The fort was financed by «Manoel» himself. Today the little island is known as Manoel Island. De Vilhena was highly popular among the Maltese and became affectionately known by his first name - Manoel. He founded the suburb of Floriana where his name is still much revered. In Valletta, he built the charming Manoel Theatre, now one of the oldest extant in Europe. He was also responsible for the other important fort built in the 18th Century - Fort Chambray on the Island of Gozo. Designed during his rule to act as a fortified town, it would stand on high ground over-seeing the channel between Malta and Gozo and overlook the main harbour of Mgarr. Because of the enormous expense of such an ambitious project, it had to be shelved until 1749, when one of the Order's heroes, Jacques François de Chambray offered to finance it privately. Fort Chambray was completed in 1761 but regretably, it never developed into a town.

Because of the continual menace of the North African corsairs, the Order's navy was always kept on the alert. During the mastership of Perellos many sea battles were fought against them. The Knights also took part in the war between Turkey and Venice which ended in 1718. The Order was always ready, willing and able to come to the assistance of the Christian powers in times of conflict with Islam. However much appreciated was this assistance, it was clear that the stronger European powers, often in conflict among themselves, became increasingly irritated by the Order's strict adherence to their definition of neutrality.

In the early half of the 18th Century, the Order of St John was already in a state of dilemma. Anxious as the Knights were to preserve their territorial sovereignty, at a time when the modern «nation state» was evolving in Europe, their policy of neutrality was becoming increasingly ambiguous. It was clear that they could never be independent of the European powers. Despite their pretensions to grandeur and power, the Knights knew that they could never hope to compete with them as equals. Their territorial base in Malta was too small. Their resources and revenues from the priories and commanderies were minute in comparison. Sometimes, their diplomacy also failed them in the new Europe. This was the epoch of the wars of succession in Europe, of secret diplomacy and of the great alliances, intrigues and counter-reversals of policy. No wonder the Order would run into difficulties in reconciling their neutral stance. Lip service would always have to be paid to France, since from France was derived the major revenue of the Order. Spain too, had to be shown respect since she ultimately controlled Sicily's destiny. Whoever controlled that island could strangle Malta at a stroke. In times of dispute, sanctions could be placed and Malta would be denied vital provisions of food and other imports. Increasingly, the Mediterranean was being dominated by the emerging economic rivalry of France and Britain, who in the pursuit of colonial trade and in the expansion of their maritime power, were now bent on a collision course. Sooner or later, therefore, the strategic value of Malta and her fine harbours would again prove to be too irresistable a prize to remain in the hands of a vacillating mediocre power. In the new power game, the Order of St John had few cards, if any, to play.

The questions of sovereignty and neutrality would arise frequently in the 18th Century. The Order's diplomacy was to become more flexible and take into account the vicissitudes of European politics. Grand Master Marc Antoine Zondadari (1720-1722), clearly unhappy with the Austrian connection of Spain, found it difficult to reconcile the Order's sovereignty over Malta with his obligation to confirm its allegiance to Spain annually, when the nominal «rent» for the Islands (the traditional presentation of one falcon) was due. Zondadari's brief mastership was marked by lean times in Malta. He was perhaps one of the few Grand Masters actually to take stock of the Order's precarious financial position. When France devalued her currency during his rule and therefore reduced the Order's revenue from its estates in that country, he immediately embarked on a crash programme of economies in Malta - a move which did not much endear him to the population.

Better times followed in the succeeding mastership of Manoel de Vilhena. Indeed, as mentioned earlier, his rule proved to be one of the most productive and popular in Malta. He was the arch-benefactor who spent lavishly out of his own pocket and who generally created a more cohesive and contented society in the Islands. However, he too, had his moments of trial over the question of neutrality. He resisted Bourbon Spain's interference in Malta and he was compelled to restrain the French from using Malta's ports as though they owned them. Similar difficulties were experienced by his successor, Raymond Despuig (1736-1741).

However, the bravest - and most arrogant - stand in defence of the Order's sovereignty over Malta would be put up by one of the most controversial of Grand Masters in Malta's history. With singular determination, Manuel Pinto de Fonseca (1741-1773) pursued this inviolable principle to its very limits. During his mastership (the longest of any Grand Master) he was prepared not only to «bluff» his way through the thick and thin of threats from the Bourbon King of the Two Sicilies but also to play on that King's differences with the Pope to his ultimate advantage. In 1754, Charles IV King of Naples and Sicily, promptly penalised Malta and imposed sanctions on her trade with Sicily after Pinto refused entry into Malta of the King's emissary. Pinto immediately made arrangements with the King of Sardinia for his vital provisions. At the same time, he opened up Malta's ports to trade with all countries. In a veiled threat, that would include Turkey. It was not till a year later that, under papal pressure, the sanctions were lifted. In 1768, after the same Charles - now enthroned as King of Spain - had driven the Jesuits out of Spain, following their expulsion also from Portugal and France, Pinto seized on the opportunity also to expel them from Malta and thus sequestrate their property in Malta. To the now furious Pope Clement XIII he explained away his action by saying that it was largely prompted by his compulsion to toe the line and follow Bourbon Spain's lead. After some haggling with Rome as to how to make best use of the Jesuits' estate, Pinto finally got his way. In 1769, he recieved Papal approval to set up a university with specific faculties for higher education in Malta. Pinto thus became the founder of the Old University of Malta which today is the oldest university in the Commonwealth outside Great Britain.

It can be said that under the enigmatic rule of Manuel Pinto de Fonseca, the Order of St John reached the zenith of its international status and recognition. Always ambitious, Pinto regarded himself equal to the reigning monarchs of Europe. His ambassadors were accredited to their courts. In essence, he was the complete embodiment of that breed of 18th Century enlightened, benevolent but «absolute» dictators. He consulted nobody and nobody dared challenge his authority- except the slaves in Malta. He hanged 60 of them when he got wind that they were about to rebel and assasinate him.

Pinto's acchievements in Malta were many. It was he who built the «new» and most impressive Auberge de Castille. However, a high price would be paid later on for his failures. His ostentation would drain the Order's reserves to the point of bankruptcy. Discontent, apathy, corruption and a general laxity would spread within the Order and among the population. Dying at the age of ninety-two, Pinto had simply reigned too long. The subsequent woes that would befall the Order would all be attributed to him. His last breath on his death bed was also to be the last gasp of the Order.

Much was expected of the next Grand Master. Indeed, the election of the Aragonese knight, Francis Ximenes de Texada (1773-1775) was initially greeted with much euphoria. However, this quickly evaporated and soon it was turned into hatred. In his bid to restore the Order's finances from the precarious state which Pinto had reduced them to, Ximenes was decidedly over-zealous and singularly untactful. The series of economic reforms which he immediately put into effect were a total disaster and a recipe for further deterioration and discontent. Public expenditure was severely curtailed. He all but closed down the University. Men were laid off and salaries were reduced. Money supply dwindled and unemployment soared. With the devaluation of Malta's currency, imports became too expensive. The price of bread shot up. Soon, the Maltese were out of food and out of work. Ximenes now decided to breed more rabbits in the hope of increasing the supply of rabbit meat. He therefore outlawed shooting altogether. This was the last straw for the Maltese. Much dissension followed and when some priests were caught hunting rabbits, relations between Ximenes and the Bishop of Malta - never good at the best of times - deteriorated rapidly. Matters came to a head in 1775 when a group of about 24 priests and clerics, led by Don Gaetano Mannarino plotted a general uprising against the Order. They managed to take Fort St Elmo and St James Cavalier in Valletta by surprise. However, the element of surprise was lost when the rest of the population failed to stage expected simultaneous rebellions elsewhere. The plot had failed miserably and the priests were rounded up. Ximenes ordered the beheading of three of the conspirators and had their heads displayed on pikes from St James' Cavalier. The rest of the rebels were either exiled or imprisoned. Mannarino ended up in chains in Fort St Angelo. Pressure against Ximenes combined with a clear hostility towards the Knights now increased. Under constant threat of assasination, Ximenes was to die in his bed within three months. One of the most ignominious of rules over Malta thus came to a timely end. But the serious rift which now existed between the Order and the Maltese would be difficult to erase. It

continued page 178

Page 179: Portrait of Grand Master Pinto by Antoine de Favray (1706-1798). National Museum of Fine Arts, Valletta.

The Armoury in Valletta contains over 5000 exhibits of arms and suits of armour.

Verdala Palace overlooking the orange groves of Buskett, not far from Rabat, Malta. It was used as a summer residence by Grand Master de Verdalle. Designed by Gerolamo Cassar in 1586.

would take a man of extraordinary ability to restore order and confidence in the tottering regime of the Order of St John in Malta.

That man arrived in the person of Emmanuel de Rohan-Polduc (1775-1797). During the twenty-two years of his mastership, he probably achieved the impossible. He not only halted the tide of discontent in Malta but, in time, he also scored a personal triumph by emerging as a highly popular figure among the population. Showing great clemency, de Rohan immediately pardoned all those involved in the «uprising of the priests». He then embarked on a vigorous programme of popular reforms which would be of lasting benefit to the Maltese. He restored the University to its former status and introduced many educational reforms. With piracy on the decline, he encouraged the growth of a Maltese mercantile fleet.[1] After a few early difficulties, the Regiment of Malta which he raised and into which the Maltese were finally recruited, proved to be a great success.[2] De Rohan also introduced important legal reforms. The Code Rohan, along with the Codex of Justinian, still forms the basis of the Common Law of Malta to this day. An intelligent, enlightened and cultured person, de Rohan kept himself informed of the many important developments then taking place in Europe. His humane approach and relative liberalism are reflected in the reforms which he initiated in the Islands. His was an open mind. It is said that he «enjoyed society and was the first Grand Master to admit ladies to his receptions».[3]

Clearly, de Rohan foresaw the enormous difficulties which lay ahead and made a determined bid to ride out the storm. The situation in France was not very encouraging. He therefore sought to cultivate better relations with other «friends» in Europe. Russia's flirtation with the Order was now of long standing. With the Mediterranean increasingly dominated by Britain and France, the Russians always kept a keen eye on Malta. Soon, de Rohan was successful in setting up a new Anglo-Bavarian Langue in the Order. Having won the support of George III and the Pope to revive the English Langue, he combined it with the new langue of Bavaria where the Order had good friends.

However, despite his statesman-like qualities and undoubted ability, de Rohan could not stem the tide of resentment against the Order within French revolutionary circles. The financial support he would so promptly give to Louis XVI, as he struggled to save himself from the guillotine, would not endear the Order to the revolutionaries in France, now engaged in «the First Terror» against royalism, feudalism and Christianity. To them, the Order was no more than a «puppet» regime supported by hostile powers. It was an anarchronism which despite its posture was neither sovereign, nor military, and most certainly, not religious any longer. It was merely «an institution to support in idleness the younger sons of certain privileged families».[4] With the birth of the First Republic in 1792, the financial life line of the Order was severed when its estates in France were all confiscated. To the horror of monarchial Europe, the excesses of the revolution continued unabated. In 1793, Louis XVI was guillotined. In the same year Marie-Antoinette followed him. The excesses were such that the Cathedral of Notre-Dame in Paris was renamed as the «Temple of Reason» and a chorus girl was there installed as the «Goddess of Reason». However, the ideals of the revolution clearly survived. In Malta, they had penetrated the ranks of the French knights who were now already undermining the Order with their subversive activities. As Napoleon Bonaparte was being glorified for the outright success of his campaign in Italy in 1797, a saddened Grand Master in Malta passed peacefully away. Destiny was kind to de Rohan. He had at least been spared from presiding over the final humiliation of the Order of St John in Malta.

If destiny was kind to de Rohan, then it can equally be said that history was most unkind to his successor, Ferdinand von Hompesch (1797-1798). It was his destiny not only to become the first German Grand Master of the Order but also to be its last in Malta. On him, history would heap all blame and shame for the downfall of the Order. Von Hompesch would be accused of having been weak-willed, cowardly, irresolute and incompetent. He would be reviled and become the victim of a torrent of abuse. In the end, he would even be sacked by the Knights themselves. His name would be forever associated with the ignominious capitulation of Malta to the French in 1798.

Although Hompesch was the convenient scapegoat for all the ills and woes of the Order, he was possibly not the arch-villian that he was made out to be. That he was not the right man to be in command during such a momentous period is certain. However, he was not an unpopular figure. He had held high posts in the Order and his contacts in central Europe were considerable and useful. He was also Maltese speaking and particularly well liked in the villages. Indeed, Zabbar took the secondary name of Citta Hompesch. He also lent his name to the village of

Siggiewi which became known as Citta Ferdinando.[5] At the same time, it was clear that he was not a dynamic person. Always hesitant, he did not inspire much confidence. With the loss of the priories in France, the treasury in Malta was exhausted. Industry and commerce were fading rapidly and the once mighty fleet was a sorry sight as it lay idle in the Grand Harbour.

For a long time, it had been clear that the odds were heavily stacked against the survival of the Order in its present form. It would have to abandon its neutral stance and enter into one alliance or another. The British, although clearly interested in Malta, were hesitant. Negotiations with Russia were at an advanced stage and Tsar Paul I had already assumed the title of «Protector of the Order». No doubt, this hastened the decision of the French Directory to order Napoleon to take possession of Malta whilst on his way to Egypt, before anyone else did. Time was not on Hompesch's side. Nor was the «fifth column» of some French knights within the Order itself - and the French contingent numbering about 200 knights clearly outnumbered all the other knights. What is strange is that Hompesch did hardly anything to remedy the perilous situation confronting him. Possibly, he was playing for time. He may also have concluded that Malta, in her uncohesive state, could never hold out against the might of the French army. Possibly, he was also hoping that, with an act of super-diplomacy, he might arrive at some kind of an accomodation with Napoleon, maintain the Order's presence in Malta and thus win some more time. But whatever his reasons, Hompesch had obviously misread the situation.

On 6th June, 1798, the formidable French fleet, consisting of over 300 vessels which included fourteen ships-of-the-line and thirty frigates, appeared off Malta. The advance guard approached the Island and sent a message requesting permission to enter and take on water. Permission was granted but it was restricted to two vessels at a time. Three days later, the main body of the fleet, under the command of Napoleon in his flagship «L'Orient» also approached Malta. Napolean now demanded water for all his fleet. Von Hompesch immediately sought a Council decision over this important issue. It was agreed that only four warships at a time should be allowed in. Napoleon considered this impractical and unacceptable. In a strong reply, he declared his intention to take by force what the Grand Master had so discourteously denied him. French troops were now landed at strategic bays in Malta and Gozo. The few isolated pockets of resistance they met with, especially in Gozo, belonged largely to the Maltese Militia. Some of the French knights merely surrendered their positions and, crossing over to the other side, now actively incited the Maltese against the Order. There was chaos and confusion in Valletta. Within thirty-six hours, the French army had overwhelmed the feeble resistance they had encountered. The Maltese, clearly disillusioned with the Order's inability to save the situation, now requested Hompesch to sue for peace. Soon, Napoleon was demanding the total surrender of the Islands. In the early hours of 12th June, 1798, on board L'Orient, the armistice was signed and delivered. The capitulation of Malta was swift and complete. Under its terms, the Islands and the Order's property would be surrendered to the French. The Grand Master would be pensioned off, as would the French knights who were given the option to return to France or remain in Malta if they so wished. The Maltese would be assured of their rights and privileges and freedom to practice their religion.

Also on 12th June, the battleship L'Orient entered the Grand Harbour. The new conqueror, General Bonaparte, hardly 29 years old, disembarked and set foot on Malta for the first time. Briskly, he climbed the steps up the steep streets leading to the Palace of the Grand Master in Valletta. With the tri-colour now hoisted over the untested fortifications and massive bastions of Valletta and elsewhere, it was clear that he was the undisputed master of the moment.

Napoleon had finally acchieved what for centuries the Turks and Islam had failed to do. He had brought the proud Order of St John to its knees. Curiously, he had subdued the impregnable fortress citadel of Valletta without having to scale its walls, escarpments and parapets, and without even having to subject them to the fire-power of his artillery. With hardly a shot fired, and with the loss of exactly three French soldiers dead, one of the most formidable and impressive concentrations of defensive systems in Europe had caved in to his advance. From its forts he took 1200 guns, 40,000 muskets and 1,500,000 tons of powder'.[6] The bluff of the supposed Sovereign and Military Order of St John had been called.

On 18th June, 1798, Grand Master Ferdinand von Hompesch and a few other loyal knights made a swift exit out of Malta. As they sailed out to sea, they were leaving behind them the great city which the Order had built and so fondly embellished - largely from the very fortunes of the aristocratic Europe which Napoleon was now seeking to dismantle. Behind them too, they left an Island and a population whose quality of life they had so enriched over the two centuries of

continued page 191

The Co-Cathedral of St John in Valletta. Designed by Gerolamo Cassar and financed by Grand Master de la Cassière, the Conventual Church of the Order was built between 1573-77.

The Oratory of St John in the Co-Cathedral in Valletta. Erected during the Mastership of Alof de Wignacourt in 1603. Behind the altar is Caravaggio's masterpiece 'The beheading of St John'.

The vault of the interior of St Johns' in Valletta. Painted in oils directly on the primed stone by Mattia Preti between 1622-66. The paintings depict 18 scenes from the life of St John.

La Valette's armorial emblems in the Palace, Valletta.

The Crypt of St Johns' in Valletta. Tomb of Grand Master Jean de la Valette is on the right.

DIMOSTRAZIONE DI TUTTE LE BATTERIE

LA FUGA, E PARTENZA DELL'ARMATA
TURCHESCA, A DI 13 SETTEMBRE 1565

Scenes from the Great Siege 1565 painted by Matteo d'Alleccio - a pupil of Michelangelo. Twelve paintings on this theme form the frieze in the Hall of St Michael and St George, formerly the Grand Council Chamber, in the Palace of the Grand Masters.

The Great Siege monument by Antonio Sciortino (1879-1947) in Republic Street, Valletta.

The fine balconade of the Magisterial Palace in Valletta.

The Knights of St John restore the walls of Jerusalem. From one of the friezes in the Palace of the Grand Masters in Valletta.

The tomb and monument to Prince Louis Charles d'Orléans in the Chapel of St Paul of the Langue of France in St John's Co-Cathedral in Valletta. He was the son of the Duke Louis Philippe Joseph, known as Philippe Egalité, and brother to King Louis Philippe of France. Prince Louis Charles died of consumption in Valletta at the age of 29 in 1808. The marble monument is the work of the French sculptor Pradier.

The Magisterial Palace and Armoury in Valletta. Gerolamo Cassar designed the Palace of the Grand Master in 1571. It is richly decorated and furnished. Originally the official residence of the Grand Masters until 1798, it was subsequently used by British Governors from 1800 onwards. It is now the residence of the President of the Republic of Malta and the seat of the House of Representatives.

The Church of Our Lady of Victory in Valletta - the oldest in the city. It was built in 1536 to commemorate the victory over the Turks.

their rule and cosmopolitan influence. Despite their present ignominy, their significant contribution towards the material and moral character of the Maltese Islands would never be forgotten by the inhabitants.

The golden era of the Sovereign Military Hospitaller Order of St John of Jerusalem, of Rhodes and of Malta had come to an abrupt end. The Order was again homeless. Although never in the limelight of international events, it would not simply fade away. Today, it still survives as a highly efficient philanthropic organization in the hospitaller tradition. From its headquarters at the Palazzo Malta in Rome, it directs its hospitals and clinics in many parts of the world and provides prompt assistance to disaster areas and to a host of under-developed countries. The aims of its founder, Brother Gerard, are now probably better served than ever before.

LIST OF GRAND MASTERS OF THE ORDER OF ST JOHN IN MALTA

Philippe Villiers de L'Isle-Adam	French	1530 - 1534
Pierre del Ponte	Italian	1534 - 1535
Didier de Saint-Jaille	French	1535 - 1536
Jean de Homedes	Spanish	1536 - 1553
Claude de la Sengle	French	1553 - 1557
Jean de la Valette-Parisot	French	1557 - 1568
Pierre del Monte	Italian	1568 - 1572
Jean L'Evêque de la Cassière	French	1572 - 1581
Hugues Loubenx de Verdala	French	1581 - 1595
Martin Garzes	Spanish	1595 - 1601
Alof de Wignacourt	French	1601 - 1622
Louis Mendez de Vasconcelles	Spanish	1622 - 1623
Antoine de Paule	French	1623 - 1636
Jean de Lascaris-Castellar	French	1636 - 1657
Martin de Redin	Spanish	1657 - 1660
Annet de Clermont-Gessant	French	1660
Raphael Cotoner	Spanish	1660 - 1663
Nicolas Cotoner	Spanish	1663 - 1680
Grégoire Carafa	Italian	1680 - 1690
Adrien de Wignacourt	French	1690 - 1697
Raymond Perelles y Roccaful	Spanish	1697 - 1720
Marc Antoine Zondadari	Italian	1720 - 1722
Antoine Manoel de Vilhena	Portuguese	1722 - 1736
Raymond Despuig	Spanish	1736 - 1741
Manuel Pinto de Fonseca	Portuguese	1741 - 1773
Francis Ximenes de Texada	Spanish	1773 - 1775
Emmanuel de Rohan-Polduc	French	1775 - 1797
Ferdinand von Hompesch	German	1797 - 1798

1. Vella: Op. Cit. Vol. II p. 184
2. From "Colonel de Freslon and the raising of de Rohan's Regiment of Malta" by Charles J. Sammut in The Armed Forces of Malta Journal No. 31, April 1979 p. 3-6
3. Sir Harry Luke: Malta An Account and an Appreciation (George G. Harrap London 1960 Second Edition) p. 111
4. Luke: Ibid: p. 111
5. De Rohan, before Hompesch, had also given his name to the village of Zebbug - Citta Rohan, where his name is still much revered.
6. Hughes: Op. Cit. p. 229

Celebration of the Quatorze Juillet In Malta, 1798. Palace Square, Valletta - the nobility laying their diplomas of the titles 'at the foot of the arbre de la liberté'. A painting by H. Zarb.

The landing of Napoleon at Malta, 12th June 1898. A contemporary engraving.

THE UPRISING AGAINST THE FRENCH

Napoleon Bonaparte stayed in Malta for exactly six days. Through a series of daily ordinances, he swiftly attempted to usher in sweeping reforms and turn the Islands into a carbon copy of a typical department of republican France. For the feudal-minded Maltese, the reforms, which Napoleon was daily promulgating, were indeed radical and far-reaching. Possibly, he was over-playing his hand. The conditional welcome he had recieved from the Maltese was largely prompted by a dislike of the Order of St John which, they felt, had betrayed them. Napoleon was clearly misinterpreting this for genuine support of the ideals of revolutionary France. He was also dangerously underestimating the sensitivity of the Maltese towards anything that smacked of tyranny or repression of their traditional rights and privileges, which were always bound up in the existing social structure; now clearly dominated by the Church and the nobility.

Although some of the reforms were attractive, particularly those leading to the establishment of primary school education, much of the new legislation was nothing more than a veiled assault on the Catholic Church and the nobility, suitably sugar-coated in the sanctimonious jargon of the new ideology. First to go was the nobility. Given that this was the aftermath of the French Revolution, this was understandable. At the same time, all traces of the past regime, in the form of armorial emblems and inscriptions on the facades of the palaces and the great houses and on public monuments, were to be removed. The names of the piazzas and streets in Valletta and elsewhere were to be substituted with ones which better conveyed the spirit of the revolution. Thus the Palace Square would become «Place de la Liberté». Present day Republic Street - then, Strada San Giorgio - would be renamed «Rue Nationale». St Christopher Street would recieve the high sounding name of «Rue des Droits de l'Homme». To drive the point home, the Maltese were also to wear the blue, red and white cockade.

Next came the curbs on the Church. All foreign clerics were to be deported. The religious orders were to be restricted to one convent each. No priests were to be recruited until existing ones were found suitable employment. The minimum age for taking holy vows by novices was to be raised to 30 years. The Pope was to have no jurisdiction over Malta and appeals to Rome were to be banned.

During his brief sojourn in Valletta, Napoleon was not only busy drafting new laws. He was also ransacking the churches and palaces and stripping them of their treasures - gold and silver items, precious stones and anything of value. All were being loaded onto his flagship L'Orient. In due course, this treasure, then valued at just under a quarter of a million pounds sterling, would sadly end up at the bottom of the Nile delta, just off Aboukir Bay, where L'Orient was destined to sink after being blown up by Nelson in the famous Battle of the Nile.

It was also clear that the wealth which the Order of St John had left behind in Malta was being used not only to prop up the unsound financial position of the French administration, but also to finance Napoleon's adventures in Egypt. It is said that he was able to pay his troops during the Egyptian campaign from the proceeds of the melted silver bullion extracted from the utensils looted from the Sacra Infermeria in Malta. So much for the principles of égalité, fraternité and liberté which were being so vigorously propagated in Malta!

Within three months of the French take-over, the Maltese were already restive. They were exasperated with the flagrant abuses of the French garrison. No longer could they tolerate the extortionate taxation, the despoiling of their churches, including that of the Cathedral of Mdina,[1] the general arrogance of the troops and the outright tyranny and despotism of the administration. When news reached Malta of the French defeat at Aboukir Bay, the Maltese prepared for a revolt. They pounced on the small French garrison in Mdina killing the commandant, François Masson, and massacring the rest of his soldiers. The rebellion quickly spread to all parts of the Islands. Soon, the French were forced to withdraw behind the main fortified areas in Valletta and the Three Cities, into Forts Ricasoli, Tigne and Manoel in Malta and into Fort Chambray and the Gran Castello in Gozo.[2] General Vaubois, the French Commander, was clearly surprised with the sudden eruption of violence by the Maltese. He later wrote - «we had to combat enraged lions, no trace of their former docile character appeared».[3]

The Maltese had, by now, organised themselves under the leadership of Emmanuele Vitale, Canon Saverio Caruana and Vincenzo Borg, also known as «Braret». In Gozo, Don Salvo Cassar was in charge. However, the newly formed militia units were too poorly armed and ill-equipped to dislodge the French by force. Several attempts to break into Valletta failed with much loss of lives. The French would have to be blockaded and starved out; but in the meantime, assistance had to be sought from elsewhere.

Appeals were made to the King of the Two Sicilies and to Nelson who, with his fleet was wearily making his way back to Naples from the Battle of the Nile. With Naples seriously menaced by an advancing French Army, King Ferdinand was clearly having troubles of his own. Material support from that quarter would be slow in coming. More prompt was Nelson's response when he recieved a Maltese deputation on board his flagship HMS Vanguard off Sicily.[4] Because his own battle fleet was badly in need of re-fitting, he requested his Portuguese allies to proceed to Malta and blockade the harbours. By October 24th, 1798, Nelson was able to relieve the Portuguese and land arms and munitions in Malta. Another consignment soon arrived from King Ferdinand. Nelson left Captain Alexander Ball of the Royal Navy in command of the blockade who, at the invitation of the Maltese, also assumed overall command of the Islands. In the meantime, Fort Chambray in Gozo had fallen into Maltese hands. With the support of the British, the Gran Castello was also overwhelmed. The British flag was thus raised for the first time in the Maltese Islands.[5]

The plight of the entrapped Maltese within the walls of Valletta was now becoming desperate. The leaders of the Maltese insurgents, and notably Braret, had always maintained contact with them. It was now agreed that a combined and co-ordinated assault on Valletta should be made. A secret plan was drawn up with the leaders of the Maltese within the city. However, on the appointed night, the French commander of Fort Manoel accidently stumbled onto the plot when he noticed undue activity on the waterfront as the Maltese insurgents were taking up their positions in preparation for the attack. He immediately gave the alarm and the plan had to be abandoned. Many Maltese suspects were later rounded up in Valletta by the French. 43 of them were shot. Amongst them were the two ringleaders and heroes of the uprising: Don Michele Scerri, a philosopher and mathematician, who was highly popular in Valletta, and Guglielmo Lorenzi, a Corsican corsair who had quite recently fought with the Russians against the Turks. Because both of them were known not to be much in favour of a British presence in Malta, it has been assumed in some quarters, but without any firm evidence, that Lorenzi was a Russian agent.[6]

With the fall of Naples to the French in November 1798, it was feared that Sicily too might soon be taken. The Maltese could not count on further support from King Ferdinand who had now fled to Palermo. In Malta, the situation was becoming increasingly desperate. The blockade had seriously disrupted the feeble economy and food was now scarce. The French were still holding out in Valletta. Maltese hopes to end the blockade and drive the French out were now being pinned on additional British re-enforcements. In late December 1798, a detachment of British Gunners arrived and soon their mortars were in action. However, the blockade at sea was not always effective and occasionally, French ships would slip through. Indeed, the British naval blockade had to be lifted in May 1799 when the Brest fleet under Admiral Bruix succeeded in entering the Mediterranean.[7] This diverted Nelson's attention from Malta for a while. It became clear that the blockade was going to be a long drawn out affair, especially now that the French could obtain provisions virtually at will. As the months rolled by, Captain Ball, now a highly respected figure in Maltese circles and recently appointed Governor of the Island by the King of Naples, continued to press for more British troops to be sent out to Malta. Nelson, who was deeply concerned, was also actively soliciting for support from King Ferdinand. It was not till December 1799 that two British battalions, under the command of Brigadier-General Thomas Graham, arrived in Malta from Messina. Assuming overall command of land operations in Malta, Graham again requested yet more soldiers. A token force of about 2000 mixed Neopolitan troops soon arrived. Not satisfied, Graham went ahead with his original plan to raise a Maltese battalion. In March 1800 he called for volunteers with his famous cry, «Brave Maltese, you have rendered yourselves interesting and conspicuous to the world. History affords no more striking example». Recruits poured in. The Maltese Light Infantry, also known as I Cacciatori Maltesi thus came into being. It would be the precursor of the subsequent long tradition of Maltese Corps in the British Army.

Largely because of the unsuccessful attempt by the combined regiments to wrest Valletta from the French in June 1800, Major-General Henry Pigot now arrived in Malta with a detachment of about 1,500 troops. In July, he relieved Graham of his command. At the same time the resumed British naval blockade was stepped up. French ships found it increasingly difficult to break through. The situation in Valleta was becoming grim. There was no food and disease was rampant among the French troops. General Vaubois, who by his stout resistance and leadership, had so clearly distinguished himself throughout the long siege, was now weakening in his resolve. His 4000 or so weary and famished garrison was being confronted by a total of 2300 British infantry, 400 Marines, a Neopolitan force of about 2000, 800 Maltese Light Infantry apart from an additional 2000 Maltese irregulars.[8] However, it was famine that finally drove him, reluctantly, to call for a negociated withdrawal from Malta.

To the lasting shame of the British, the Maltese, who were justifiably proud of having originally pinned the French down and who had voluntarily sought assistance from the British, were excluded from taking part in the negotiations. Ball, the champion of the Maltese cause protested to Pigot. Relations between the two were never good at the best of times. Pigot maintained he was acting under instructions from General Sir Ralph Abercromby, Supreme Commander of British military forces in the Mediterranean. Pigot was adamant that not only would the Maltese not be represented but, once Valletta was in British hands, only the British flag would be hoisted; despite the fact that the Maltese had already raised the Neopolitan flag in the areas which they controlled in both Malta and Gozo. Ball himself, was excluded from the discussions with Vaubois, who resolutely demanded that the Maltese must be prevented from entering Valletta until the French evacuation was complete. Graham's attempt at least to allow the Maltese to be present for the signature of the capitulation which took place on 5th September, 1800, also came to nothing. Three days later, the French were out of Valletta and most of them were on their way to Marseilles.

The British fleet now entered the Grand Harbour. For the next century and a half, it would there make its home in the Mediterranean. The last British warship, HMS London, would not leave the Maltese Islands till April 1st, 1979.

1. Vella: Op. Cit. p. 225
2. A. Samut-Tagliaferro: History of the Royal Malta Artillery, Vol. I. (Lux Press Malta 1976) p. 9
3. Samut-Tagliaferro: Ibid. p. 9
4. A.V. Laferla: British Malta Vol. I (A.C. Aquilina & Co, Malta 1945) p. I
5. The British flag was replaced by that of the King of Naples at the insistance of Don Salvo Cassar.
6. Vella: Ibid. p. 236
7. Laferla: Ibid. p. VI
8. Samut-Tagliaferro: Ibid. p. 17

THE BRITISH PERIOD

Autocracy to Democracy and back to Direct Rule

The early beginnings of British rule in Malta were shrouded in uncertainty. Having achieved naval supremacy in the Mediterranean largely as a result of Nelson's resounding victory over the French in the Battle of the Nile, Britain had stumbled into Malta almost by accident. She had come at Malta's request, to assist the Islanders in driving out the French. Promptly, Britain had rallied to Malta's cause which, ultimately, was also her own. Mission accomplished, the British now found themselves sovereign in the land. Having always regarded Malta's sovereignty as the legitimate preserve of her ally, the King of the Two Sicilies, Britain was at first reluctant to hold on to the Islands. Indeed, when the Maltese had rebelled against French rule, it was the Neopolitan flag which they had raised. True, this was done largely because they feared that their own would be unrecognized. It was also true that the Maltese had raised no objections when the British flag was hoisted over the battlements of Malta. On the contrary, they seemed to be delighted, judging by the sentiments of solidarity and loyalty they were now expressing in favour of the Crown. It was all very bewildering, and even embarassing. Unlike France, Britain had yet to be convinced of Malta's strategic value. As long as France would never regain the Islands, and as long as their neutrality could be guaranteed, Britain was not averse to their being returned to the Order of St. John, now that the Emperor of Russia was the 'de facto' Grand Master. Maltese unacceptance of such a situation would be of no consequence. Their declarations of allegiance to the British Crown would be equally irrelevant.

In 1802, the Treaty of Amiens brought an uneasy peace between Britain and France. For Malta, it was a foretaste of Britain's open disregard of Maltese opinion, which would be a source of major irritation for decades to come. In line with British thinking and much to the satisfaction of Napoleon, Article X of the Treaty clearly stated that Malta was to be returned to the Order and placed under the 'dubious' protection of Naples. Britain, France, Austria, Russia, Spain and Prussia would guarantee her neutrality. Unhappy with this absurd arrangement, the Maltese felt that they were back to square one. They immediately drew up a Declaration of Human Rights, which among other things, asserted that they regarded their Sovereign to be the King of the United Kingdom of Great Britain and Ireland and that Britain had no right to cede the Islands to any other power. That right belonged to the inhabitants. Malta would never return to the Order or again form part of the Kingdom of the Two Sicilies.

Fortunately for Malta, the so-called 'peace' of Amiens was anything but peaceful. There were violations of the Treaty on both sides. Britain was soon having second thoughts about Malta. With arrangements well in hand for the Order to take over, and with Napoleon seemingly so anxious to secure the evacuation of the British from the Islands, Britain reversed her decision. A furious Napoleon was to remark to the British Ambassador in Paris, "Peace or war depends on Malta. I would rather put you in possession of the Faubourg Saint Antoine than in Malta. " By March 1803, he had given Britain an ultimatum to quit the Islands. Two months later, war was again declared, Clearly, Britain had now decided to keep Malta.

With the resumption of the Napoleonic wars, the British came to realise the value of Malta, not only to control Southern Italy and the Levant, but also, as Nelson put it, as a 'most important outwork to India'. Increasingly, it would be from Malta that Britain would be able to check French ambitions in the Mediterreanean and also keep tab of the worsening situation in Greece and in the Balkans. Britain's naval superiority, considerably boosted in 1805 with the defeat of the French fleet in the Battle of Trafalgar, would become the keystone of her imperial strategy. Malta would become part and parcel of that grand design.

At first, Britain sought to maintain the 'status quo' in Malta. The Islands would be under her protection. The rights, privileges, customs and religion of the population would be respected. The Civil Commissioner, Charles Cameron, would ensure that good will towards Britain was maintained. With the collapse of the Treaty of Amiens, and with British intentions somewhat clearer, Britain now started to assume more responsibility for the Islands' affairs. Strategic and imperial considerations would come foremost. Following the evacuation of the French, the series of interim administrative arrangements had led to much confusion. The responsibility to bring order largely fell on Sir Alexander Ball, now appointed Civil Commissioner. During his eight years in office, he successfully reorganized Malta and promoted her advantages as a centre for commerce and trade. He cultivated good relations with the Church, effected some important reforms, particularly in the Courts of Appeal, and generally pursued British interests without undue acrimony. Although he had to contend with the increasing agitation by the nobility for the establishment of a Maltese assembly or Consiglio Popolare, he managed to remain popular with the Maltese.

Malta's role in the British scheme of things assumed greater importance. From Malta, the approach to Egypt could be safeguarded and from here, Britain could swiftly intervene in Italy when necessary. Maltese soldiers were thus seen fighting in Calabria and in the Island of Capri, against the armies of Joachim Murat, in 1808. Events in Europe were followed with the deepest interest in Malta. Napoleon's successes at Austerlitz in 1805, Jena in 1806, Eylau in 1807 and Wagram in 1809 did not exclude the possibility of Malta being restored to France. However, Nelson's victory off Cape Trafalgar in 1805, Wellington's successes in the Spanish peninsula from 1808 to 1813, Napoleon's disastrous campaign in Russia which had cost him half a million lives in 1812, his overthrow by Wellington and Blucher at Waterloo in 1815 and his surrender at Rochefort to his greatest enemy on board the 'Bellerophon', rendered the reaquisition of Malta impossible. Under Article VII of the Treaty of Paris of 1814, Britain's sovereignty over the Maltese Islands was confirmed. It was sealed a year later at the Congress of Vienna.

The war in Europe in the first decade of the 19th Century gave the economy of Malta a welcome boost. Exploiting the situation to its fullest advantage, Ball succeded in turning Malta into a centre for entrepôt trade. This led to much activity in the harbours and to bulging warehouses. Regrettably, the 'mini-boom' would be short-lived. With the Treaty of Paris, Malta's superficial advantages disappeared.

Maltese aspirations for a larger say in the affairs of state would not disappear. The Declaration of Human Rights of 1802 had already expressed the clear intention of the nobility and the landed gentry to revive the old Consiglio Popolare. While these factions of Maltese society had clear vested interests in this proposal, it is fair to say that the Maltese conception of liberty implied that Malta should be allowed to manage its own affairs and control legislation on a footing of complete political equality. It was difficult for the Maltese to understand that the evolution of their political institutions had to be subjugated to the exigencies of imperial strategy.

They felt that the benevolent autocracy of British rule in Malta betrayed a reluctance to contemplate the beginnings of self-government in Malta. Mere protection and patronization were not enough. The Maltese were after legislative power.

In 1811, the Marquis Niccoló Testaferrata, the champion of the Maltese nobility, strongly petitioned the Crown for the revival of the Consiglio with deputies to the King-in-Council. Largely as a result of his energy and eloquence, a Commission was set up to review colonial policy in Malta. However, its recommendations were most disappointing. It quashed the argument for the Consiglio. It combined the duties of the Civil Commissioner and the military commander, now to be performed by a Governor, who would be empowered to appoint a consultative council. No portion of political power would be vested to the Maltese who were considered unfit to exercise it. Strategic considerations and imperial interests could not be placed at risk by yielding a share of Malta's autonomy to a bunch of erratic and fanatical noblemen. In total, it was a flat no to any of Testaferrata's proposals.

The value which Britain attached to Malta was spelt out in the Secretary of State's memorandum to Sir Thomas Maitland, the first Governor of Malta. In 1813 he wrote, 'As a military outpost, as a naval arsenal, as a secure place of depot for the British merchants, there is no place in the South of Europe which appears as well calculated to fix the influence and extend the interests of Great Britain as the Island of Malta.' Arriving in Malta in the midst of raging plague in 1813, Maitland resorted to drastic quarantine measures to stamp it out ; although it still claimed as many as 4500 lives. The new Governor proved to be highly efficient and effective. His reforms were many. He abolished the grain monopoly, introduced English for official correspondence, centralised public expenditure and reformed land ownership. His major contribution was perhaps in the field of judicial reform. Despite the clear absolutism of his rule, which earned him the nick-name of 'King Tom', his untimely death in 1824 was much regretted in Malta. Riding rough-shod over local liberty, he had genuinely fostered the well-being of the population.

However enlightened and benevolent were the rules of 'King Tom' and his immediate successors, they did not stem the increasing agitation among the nobility and the landed gentry for native participation in the Islands' affairs. The implied patronage of one-man rule was an insult to them. They recalled that the Maltese had, after all, placed themselves under British rule out of their own volition. Sweeping reforms were called for. Time and again, protests would be made ; but they would largely be passed over. Even the good relations with the Church, which Ball and Maitland had so successfully cultivated, would receive the occasional bruise.

But times were changing. In the late Eighteen-Twenties, a wave of liberalism penetrated the corridors of power in England. It culminated in the Reform Bill of 1832. A breath of fresh air now entered into the 'hawkish' atmosphere of the chambers of colonial policy. Two Maltese aristocrats, Mitrovich and Sceberras, quickly seized on this opportunity to make friends and gain support for the Maltese cause. As a result of their efforts, a Council of Government was set up in Malta in 1835. Consisting of seven official appointed members, the new Council was a far cry from the Mitrovich proposals ; but it was, at least, a beginning in the long road to representative government. That road would be often obstructed, on the one hand, by British intransigence to dilute the powers of the Governor and thus subordinate British interests and influence to the wayward whims of the Maltese upper class, and on the other hand, by the combative but often short-sighted spirit of non-cooperation of Maltese leaders. The over-zealous and often intolerent attitude of the clergy, suspicious to the point of bigotry of creeping Protestantism and always highly sensitive to any measure which might weaken the overwhelming influence of the Church, would also contribute to the general disharmony of Maltese politics.

The workings of the unrestricted Council over the years were never very impressive. There were many skirmishes with the Church, particularly over ecclesiastical appointments. Fearing the expression of anti-clerical opinion, the Church had initially opposed the introduction of a free press. It was finally allowed in 1839 ; but no one could publicly attack the Church in print. It was during that year that the Dowager Queen Adelaide spent three months in Malta. Shocked that British protestants had no place of worship in Malta, she laid the foundation stone of St Paul's Anglican Cathedral in Valletta. Designed by an admiralty architect, William Scamp, it was built at her own expense. However, religious bigotry was not always confined to one side. In 1846, the Governor, Sir Patrick Stuart, himself a rather zealous Anglican, caused riots in Valletta when he foolishly banned the celebration of the traditional Carnival on the Sunday before Ash Wednesday.

It was possibly because of Stuart's many provocations and known bias against the Catholic Church that the Colonial Office replaced him with a Roman Catholic Governor in 1847. A further experiment was tried. Unlike previous military governors, Richard More O'Farrell was a civilian. It was during his term of office that, after continued lobbying by Mitrovich and Sceberras, Malta received a new constitution in 1849. For the first time, there would be eight 'elected' members on the Council of Government, along with ten official members, five of whom would be Maltese. In the election that followed, three members from the clergy were returned. They would cause much dissension when the Council came to debate reforms in the criminal codes. They insisted on heavier punishments for offenders against the 'dominant' Catholic Church than for those who offended other 'tolerated' religions. Try as he might, poor O'Farrell never quite managed to reduce the incidence of religious arguments from cropping up and he would always be dogged by religious bigotry. When he sought to restrict the flow of Italian refugees into Malta, after the Revolution of Rome in 1848, he was accused of being anti-Jesuit.

In 1857, when Sir William Reid was Governor, a ready excuse was found to exclude clerics from the Council of Government. Since the Archbishop could stand in the way of a cleric wishing to stand for election, priests were not considered to be 'free agents' any longer in 'Representative Government'. A few months earlier, members of the judiciary were similarly excluded from the Council. Considering that the Church was a principal landowner in the Islands, the exclusion of her representatives was not altogether just. In time, this issue would become the subject of a referendum and priests would be re-admitted.

Despite the slow progress in the field of constitutional reform. Malta moved ahead in many other fields and particularly in those, which like defence and imperial strategy, were outside the strictures of local opinion. Increasingly, the value of Malta's strategic position became the directing force behind her future development. First strategic use of her was made in 1827. Malta benefited much from the arrival into her harbours of the combined fleets of Britain, France and Russia in preparation for their successful encounter with the Turco-Egyptian fleet which culminated in the Battle of Navarino. Malta's value was brought more into focus in the Thirties as French influence penetrated North Africa and Egypt. Malta benefited from increased defence spending in the Islands, as the foundations of a naval base were being laid. The fortifications in Malta were strengthened, the naval hospital at Bighi came into being and underground fosses, for the storage of grain, were excavated at Floriana. The requirements of the Royal Navy took priority over anything else. The small shipyards which had been inherited from the Order of St John were greatly expanded. In 1848, the first drydock in Galley Creek was opened. By 1871, Somerset dock would also be complete. In time, a total of five drydocks would be built. Harbourside facilities would be developed alongside them. Facilities for commercial shipping too would be improved and extended in the Grand Harbour. From all these developments, Malta would prosper.

The war of the Crimea in 1854-56 again brought considerable military activity to the Island. Acting as a rear base for operations in the East, it was from Malta that troops would embark for their battle destinations. Here, the wounded would be returned and receive hospital care. Malta's importance as a supply station and as a refit naval base became unquestionable.

As a commercial outpost too, Malta forged ahead. With steam replacing sail at sea, Malta became an important coaling station with an abundance of harbourside bunkers and facilities for shipping of every kind. With the opening of the Suez Canal in 1869, Malta really came into her own. She was now on the new highway to the East and to India. She therefore became the favourite port of call for all shipping plying that route and the leading trans-shipment centre for cargo and mail. With ships of every flag calling at Malta, the Grand Harbour became a hive of activity, from which everyone was to benefit. Valletta assumed a cosmopolitan character as she basked in the newly found affluence of the times.

As usual, the Islands' prosperity was quickly reflected in a dramatic rise in the population. This would continue well into the 20th Century. From a figure of 114,000 in 1842, the population rose to 124,000 in 1851. By 1871, it would reach 140,000 and it would more than double by the advent of the Second World War. With each increase, the problems of congestion, especially in the urban areas of Valletta and the Three Cities, would again become serious. Attempts would be made to encourage people to move to the newer suburbs and settlements which were sprouting everywhere. The older and more successful villages too would expand and even link up with others adjacent to them. Despite the prosperity, employment for the increasing work force would not always be available. Emigration schemes would therefore be introduced, but they would be largely unsuccessful at first. It would not be till the end of the century when, with the trade boom on the decline and Malta's fortunes ebbing, the Maltese would start to emigrate; mainly to North Africa.

With it, prosperity brought in better living standards and an increased demand for goods and services. Malta's infra-structure had to be developed to cope with each new situation. The pressures for better social services would be great. There would be problems over water supply and much hassle and petty obstructions over drainage and sewage projects. In 1881, the distillation of sea water became a necessity. A year later, electric street lighting was introduced. In line with the times, a railway line was laid between Mdina and Valletta in 1883. But this would become obsolete and disappear with the advent of electric tramways, themselves to be eventually overtaken by the motor bus.

It has to be remembered that, from the latter half of the 19th Century onwards. Malta was not just an important base for the Royal Navy. It was also a garrison station. In 1889, the garrison strength stood at over 10,000 soldiers. [1] By 1904, it would increase to 17,000. In 1929, there would be eighty-six naval vessels anchored at Marsamxett and the Grand Harbour, and over 21,000 troops in barracks in Malta. The formidable defence establishment in the Islands had to be housed, supplied and supported by an equally formidable array of ancillary services, equipment, storage depots, munition dumps and so on. The advance of technology would necessitate changes in military concepts of defence. New fortresses for better coastal defence would be built. All would be suitably armed with the latest weaponry. As iron replaced timber at sea, guns would get bigger and bigger.

The number of civil and defence projects undertaken are too great to mention. Suffice it to say, that for nearly a hundred years, the limestone quarries on Malta had to be worked at full stretch to cope with the demand for all the building that went on. But Victorian architecture in Malta was no match for the finer building that preceded it. Indeed, during the entire period of British rule, very few buildings worthy of note were erected. Apart from St Paul's Anglican Cathedral in Valletta, one could possibly only single out the Royal Opera House, also in Valletta, and Mosta Church ; both of them massive structures and both, curiously quite over-proportioned. Completed in 1866, the opera house was the most impressive. It was designed by Edward Barry who was also the architect of the Royal Opera House, Covent Garden. Regretably, it was gutted by fire in 1873, but restored to its original magnificence three years later. However, it was a fated building. In 1942, a German bomb scored a direct hit and totally demolished it. It was a sad loss. Sadder still was the fact that it has never been rebuilt. Where its auditorium stood is now a car park, in the midst of its existing ruins. The sound of screeching tyres and car horns is a far cry from the graceful arias of Bellini's 'I Puritani' which were heard on its inaugural night in 1866. The Church of St Mary in Mosta managed to survive the war ; but only just. It too, was hit by three bombs in 1942, but, miraculously, none exploded. One actually penetrated the dome and bounced on to the floor harmlessly. Built over an earlier church and completed in 1860, it was designed by a Maltese architect Giorgio Grognet de Vassé. Its dome is the fourth largest in the world.

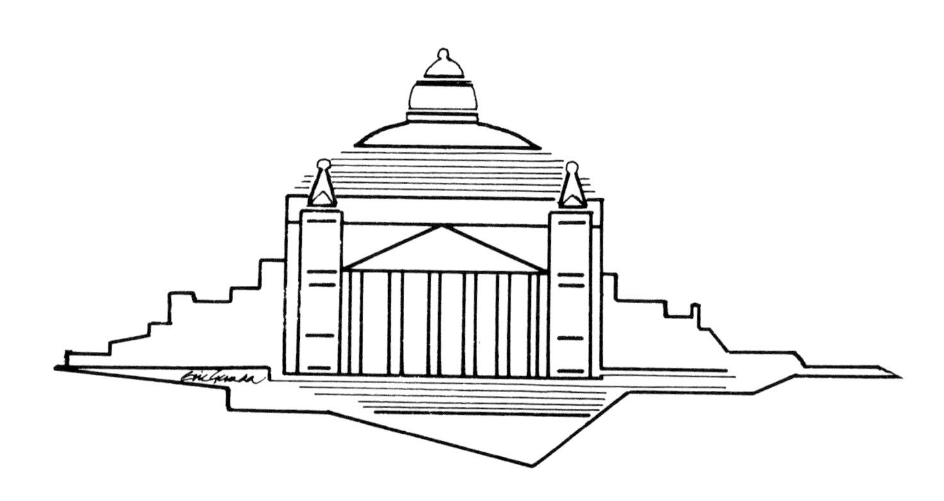

It is a curious fact that while the British enjoyed much good will and even popularity during their period in Malta, Anglo-Maltese relations, at the level of day to day internal politics, were often sour, bitter, tempestuous and sometimes, downright unpleasant. While the civil service meekly played to the tune, the conflicts between Maltese leaders and their colonial masters never ceased. To the foreigner, this was often confusing. While Maltese fidelity to the Crown and indeed, to the Empire in its heyday, was never in question, the invidious goings-on in the Council of Government were sometimes unworthy of the best of intentions of both parties. As the Maltese would rally to each and every crisis which involved or threatened British Imperial interests, they would equally resist subservience by implication, to those very interests. In the final analysis, the story of most colonial governments is much the same. Like most British colonial subjects, the Maltese went through the same traumas and, possibly like them, they developed a love-hate relationship with their masters, which could be as generous as it could be hostile. Where the Maltese were peculiarly different lay in the fact that, unlike most other colonial subjects they were essentially Mediterranean, and therefore steeped in the accrued wisdom and ancient lore of that great sea ; essentially European, with a good six centuries of Latin culture behind them, with all the virtues and failings that that entailed ; and essentially Christian and passionately Roman Catholic, with a long tradition of ritual and a good measure of self-righteousness.

Malta was no virgin territory, ripe for the infusion of Anglo-Saxon culture. She was already an amalgam of many civilizations by the time Britain entered the Mediterranean. Malta's early links with Sicily and her close association with the international Europeanism of the Knights of St John, had placed her within the cultural and religious milieu of her immediate neighbour; Italy. She was therefore more conversant with the highly developed feudal structures and institutions of the Italian City States and most willingly subservient to the overwhelming influence of Papal Rome. Industrial and Protestant Britain, emerging as an imperial power, found it difficult to reconcile her more progressive standards with those of a declining culture, of which she had little or no experience. Like most Europeans, the Maltese had long admired the relative efficiency of the Anglo-Saxon and Britain's clear lead in the advance of parliamentary democracy, emancipation and the art of government. Voluntarily, the Maltese had ceded themselves to Britain in the hope that she would immediatly introduce her much admired political institutions into the Islands on a power-sharing basis. Britain found Malta's ancient feudalism alien to her ways and not in consonance with her imperial strategy. She therefore 'patronized' Malta with a benevolent autocracy and slowly set out to anglicize the Maltese. She never quite understood or appreciated Maltese sensitivity to any proposed changes in her cultural make-up. She sometimes mistook grievance for malice and mischief which, if allowed free rein , could endanger her imperial interests. Within the imperial design and its primitive constitutional machinery, Britain was at a loss to find a comfortable and safe slot for Malta to fit into. Malta would have to inch her way in slowly to acquire those very political institutions which Britain stood for, but which she was most unwilling to impart to her colonial possessions. The consequent acrimony which marked Malta's path to self-government and ultimate constitutional independence should therefore be viewed in the light of Maltese sensitivity to the anglicization which Britain tried to impose, as the means of safe-guarding her ultimate interests. British colonial policy could not but be dominated by imperial strategy. The struggle for the unification of Italy, which Britain largely supported, the later intrigues and implications of the Triple Alliance of Germany, Austria and Italy against the Dual-Alliance of France and Russia, Italy's accord with Britain and subsequent disenchantment as a result of colonial reverses in Africa, King Edward VII's visit to Malta in 1903, his diplomatic successes in appeasing Italian wounded pride over Malta and North Africa, his courtship of the French and the subsequent isolation of Germany were all factors which weighed heavily in the vicissitudes of Malta's constitutional struggle.

Within the Malta Council of Government, the built-in majority of 'official' members over the 'elected' representatives, coupled with the ability of the Government to legislate willy-nilly through Orders-in-Council from the Secretary of State, often caused frustration in the elected camp. This would be expressed by increasing non-cooperation with the Government and incessant opposition to most of its projects, however necessary or desirable they might be. There was much opposition to the much needed drainage and sewage projects for the Three Cities where the clear lack of sanitation was a proven hazard to health. Proposals for changes in the tax system and the abolition of the bread tax also led to riots by that section of the population which would have most benefited from them - the peasant class, clearly under the influence of the upper classes and the lower clergy. Within this maze of often contradictory opinion, pressure groups were thus formed.

The struggle for the unification of Italy had brought many refugees to Malta. With their revolutionary zeal, they had introduced a fresh spirit of nationalism into Maltese political life. At the same time, the Church in Malta would express concern for the safety of the Pope and disapprove of British support of the Risorgimento in Italy. The pro-Italian faction was often split. One side would be offended by British policy towards Italy, the other would seek to imitate the new nationalism which the British were trying to repress in Malta.

The efforts of the Maltese Crown Advocate, Adriano Dingli, to harmonize Roman civil codes with those of Justinian, de Rohan and Napoleon did bring some order in the field of legal and judicial reform. However, the judicial codification of civil laws would often fall foul of the clergy. The respective interests of church and state were hard to reconcile. Progress in the field of education, largely in the hands of the Church, was equally slow. Ultimately, the issue would sadly degenerate into a battle for language supremacy between Italian and English. In turn, this would dominate Maltese political life.

When the establishment of government schools was first proposed, the lower clergy was reluctant to relinquish its control and heavy influence over primary education in Malta. Despite the erratic goings-on in the Council of Government, there still existed an element of objective opinion which not only welcomed education reforms, but also favoured the tuition of the English language, since by its dissemination, the Maltese would ultimately be in a better position to argue in favour of self-government. The chief exponent of this viewpoint was Sigismondo Savona. When in 1880, the Keenan report on education in Malta proposed giving the English language more prominence in schools, the Archbishop did not raise any serious objections, despite the lower clergy's traditional suspicions that English would only serve as a vehicle to introduce Protestantism and to arouse anti-Papal views. Violently opposing Savona, now appointed Director of Education, was the Maltese lawyer, Fortunato Mizzi, a leading figure in the struggle for a greater share in Government and whose power-base stemmed largely from the Italianate upper classes. As such he represented the pro-Italian culture groupings which were then clearly dominant in society. The imposition of the English language on the mainstream of Maltese life was anathema to them. Mizzi and the anti-reform members formed the majority on the elected Council. They would stage many a walk-out on this and other issues ; but in their absence the official members would vote in the legislation. To cripple the Council, Mizzi would successfully harness hostility towards the English language within the lower clergy, resurrect the 'Protestants under the bed' scare, champion the Italian language as the keeper of the true faith and generally rally the professional and middle class to his great cause - popular demand for total control of the Council of Government. Indiscriminate obstruction to Council business became the order of the day. The clamour for self-government would gain in momentum. By 1886, it would reach boiling point. Something would have to be done to appease Maltese demands and thus secure their cooperation in Government.

It was in that year that the young Cambridge under-graduate Gerald Strickland commenced a series of articles in the press calling for the restoration of the old Consiglio Popolare of bygone days. As a result, he was catapulted into the limelight and drawn into the fold of the faithful seeking responsible government. Strickland, of Anglo-Maltese parentage, joined Mizzi and together they presented proposals to the Secretary of State in London. But the answer was in the negative, on the grounds that responsible government in Malta would be 'incompatible with the position of Malta as an imperial fortress and unsuitable to the circumstances of the Island and the political capacity of the people'. At the same time, it was recognized that reforms in Malta were necessary and that the existing constitutional 'strait-jacket' would have to be liberalized. As Malta celebrated Queen Victoria's golden jubilee in 1887 with singular enthusiasm, despite a raging cholera epidemic, there was much haggling over the new Constitution. In December of that year, a compromise was finally reached. Malta would have majority representative Government with a Council of fourteen elected representatives and six official members under the Presidency of the Governor. There would also be an Executive Council consisting of a number of official Crown members and three appointed members from the elected representatives on the Council of Government. However the new Constitution would restrict the Council of Government's power to fiscal and monetary matters only. Real power would remain in the hands of the Governor who could always intervene and legislate by obtaining Orders-in-Council from the Crown via the Secretary of State. The Governor's power of veto would thus always hang over the elected Council. The role of the Executive Council would, in effect, be more 'responsible' than the Council of Government, which would merely exist to approve or block monies for the Government's programme.

The Constitution of 1887 could never really work. Immediately, it was clear that the Government could do what it liked, since it was not responsible to the elected council. Appointments to the Executive Council thus assumed more importance as time went by. At first, all would go marginally well, despite the immediate intentions of Mizzi to harass the Government and press for the next logical progression in constitutional development: responsible government. However, his tactics to maintain control over the elected council soon led to rebellion within his own camp and caused his temporary withdrawal from the political scene. The rising star of the show was now Savona who, returning to politics with a vengeance, started to lambast the Government with every weapon at his disposal, in pursuit of the objective of responsible government. Now friendly with the Church, he was assisted by an ecclesiastic, Mgr Panzavecchia. Together they would block votes for this and that, cause absenteeism from the Council and even secure resignations from that of the Executive. Language and religious issues again proved to be stumbling blocks, despite Savona's earlier stand on these very issues. But by 1897, Savona was a spent force and Mizzi again emerged as a political force. By this time too, the growing interest in the English language in Malta was causing many a headache to the Church and to Mizzi's Italianate faction. Amendments to the criminal laws had permitted the use of the English language in the courts. Reforms in education also gave an additional boost to that language over Italian in the schools. In the pursuit of their political aim to embarass the Government and thus put pressure for their demands for responsible government, Maltese leaders resorted to blocking votes for better social services, regardless of their dire urgency. Many walkouts from Council meetings were also staged.

At this point, it has to be said that there were other factors besides the language and religious issues which were contributing to the general political chaos in Malta. The hardening of opposing viewpoints had something to do with the strong personalities on both sides. Bitter was the antagonism which developed between the two ex-partners of the Constitution of 1887: Mizzi and Strickland. From the time of Strickland's appointment to the exalted position of Chief

Secretary in November 1888, relations between him and other Maltese politicians steadily deteriorated. Now squarely on the side of the Government, Strickland saw himself as the 'de facto' Prime Minister. His high-handed manner and uncompromising attitude led to many clashes and 'walk-outs' by Council members. Indeed, a stalemate would be reached and he would be blamed for the increasing non-cooperation of the elected Council. By the turn of the century, he had become an embarrassment to the Government and the Secretary of State considered him an obstacle to political harmony in Malta. Recognizing Strickland's undisputed ability and promise, he 'promoted' him to become Governor of the Leeward Islands in 1902. His absence from Malta during these crucial times was thus secured.

The strong personality of the Secretary of State himself, Chamberlain, also caused much dissension in Maltese political circles and often, his name was reviled in political rallies organized by Mizzi's anti-reform group. The hand of Joseph Chamberlain would act swiftly to thwart any progress towards responsible government which he saw as a threat to British Imperial interests, and particularly to the role of fortress Malta in the uncertain situation in the European balance of power. Chamberlain never minced words or hesitated to express his rather fixed views on British imperialism, the supremacy of the British race and the role of the English language in the concept of empire building. Equally, he never underestimated the influence in Malta which the substantial number of 'exiled' Italians were exerting in local politics. To him, fortress Malta was too important to 'remain' within the Italian cultural sphere of influence.

In mainland Italy, there was increasing sensitivity to Chamberlain's clear and declared intentions to 'anglicize' the Maltese. This often caused him much embarrassment. With Britain relatively isolated in Europe and under increasing threat from France and Russia, Italy's general unease, within the Triple Alliance with Germany and Austria, was of great concern to Chamberlain. For this reason, he first sought to quell Italian suspicions over Malta bu withdrawing certain extreme proclamations on the advancement of the English language in the Islands which he had earlier made. Having done that, he then asserted most firmly that the Government would not tolerate the continued blocking of funds required for its programme by the elected Council. If that continued, he would be compelled to revoke the present Constitution.

In view of this uneasy compromise, a relative calm prevailed for a while in Malta. However, Chamberlain was always ready to pounce if that spirit was violated in any way. But soon, Mizzi and his followers were again restive. Maltese grieviances about the earlier expulsion of clerics from the elected bench were again aired. The language issue again erupted. Again, the elected members commenced the blocking of votes to do with education and the teaching of English. A stalemate was reached just at the time of King Edward VII's visit to Malta in 1903; the first by a reigning British monarch. In the midst of the festivities to mark the occasion, Chamberlain was scheming a final crack down. He first waited for the Anglo-French-Italian accord to develop satisfactorily. Italy would then be reluctant to raise any protests over internal developments in Malta. By early summer 1903, Chamberlain was ready. On 25th June at that year, he struck. The Maltese were to be denied majority representative government and the clock would be turned back to 1849. Malta's new constitution would revert back to those days. The Council of Government would consist of only eight elected members against nine official ones. Moreover, no elected members would be appointed to the Executive Council.

Under the new Constitution, Maltese politicians wandered aimlessly into alternating periods of non-cooperation and absenteeism from the Council to mild concilation with the Government - in the hope of winning back seats on the Executive Council and hastening the restoration of representative government.

In 1905, following the death of Fortunato Mizzi, Francesco Azzopardi took over the mantle of leadership of the anti-reform group. It was largely due to the spirit of conciliation which he advocated, that a brief period of calm was secured in Government. It would, however, also lead to his eventual downfall within the 'Consiglio Popolare', which he had created. In time, its membership would again start preaching the doctrine of absenteeism. The British would not be entirely blameless. The Government's earlier provocation in allowing a Protestant minister to conduct a series of conferences in the Theatre Royal in Valletta had clearly re-opened old religious wounds at a most inappropriate moment. It had led to firm protests by the Archbishop, followed by a crescendo of agitated voices from the lower clergy. The usual charges against the British were tabled: deliberate undermining of the Church's privileges; denigration of the role of the Italian language in Malta's culture and religion. Indeed, with Azzopardi's departure from the Consiglio in 1911, these very charges formed the platform of Mgr Panzavecchia's 'Comitato Pattriottica' which contested the next election and obtained an outright majority on the elected bench.

The political impasse in Malta was increasingly overshadowed by the economic gloom that hung over the Islands. The position had been steadily deteriorating for a long time. Despite significant improvements in the harbour and dockside areas, which included the building of the breakwater in the Grand Harbour, Malta was facing increasing competition from other well equipped ports in the Mediterranean. Government revenue from the diminished activity in Malta's ports was falling steeply. Despite increased defence expenditure, unemployment soared. Already, it was clear that Malta's dependence on Britain's military spending was a severe handicap. If defence cut-backs were effected, as indeed they would be, much hardship would ensue. The fact that Malta had no proper system of taxation further contributed to Government's inability to balance its accounts. The more positive step to abolish the £ 5000 contribution, which the civil administration made annually in favour of the military set-up to compensate for the costs of the garrisons in Malta, was at least a sensible measure.

The winds of change in Europe and the gathering clouds of war also weighed heavily over Malta. However, political dissatisfaction never hindered the Maltese from making their contribution to the total war effort. It was in this tense atmosphere that Mizzis' son, Enrico made his first appearance on the elected bench. Educated in Italy, young Mizzi pursued the 'Italianism' of his father with much greater vigour but, possibly, with less tact. Soon, he was charged with sedition; but the Governor reduced punishment to a severe reprimand. With Malta on a war footing, the population rallied to the allied cause in World War 1. The naval dockyards again came into their own. But Malta's greatest contribution lay in the field of hospital care for the thousands of sick and wounded soldiers. For this, she earned the nickname of 'nurse of the Mediterranean'. At one stage, there were more than 25,000 beds available, some of which were even in private homes.

By the close of the First World War, Malta had to again face reality. Her wartime economy had not been unbeneficial. But with Britain's return to a peacetime role, there were to be severe cut-backs in defence spending in Malta - the chief mainstay of the economy. Much hardship and distress followed, as men were discharged from the army and from naval establishments in the Islands. Unemployment again soared and inflation ate its way into the already miserable pay packets. In spite of Government subsidies, the price of bread became too high. There were strikes and protests against low wages and the high cost of living. Increasing discontent and dissatisfaction with this sad state of affairs re-enforced the demands of Maltese politicians for progress in the field of constitutional reform, which had been held in obeyance during the war years. In a bid to achieve some unity in the morass of political viewpoints in Malta, Dr Filippo Sciberras now organized a 'National Assembly' to co-ordinate Maltese proposals for a new constitution. A huge and angry crowd gathered in Valletta for one of the meetings of the assembly on 7th June, 1919. The pent-up frustations of the people suddenly exploded into a riot. The mob soon got out of control, causing much damage to public and private property. Troops were called in to restore order; but they shamelessly opened fire. Four Maltese were killed. It took many days for Valletta to settle back to normality. Malta remained in a state of shock as a result of the unnecessary killings. To this day, there exists much controversy over this whole unsavoury episode in Maltese history.

Lord Plumer, the newly-arrived Governor had a calming effect on the explosive situation. He hastened to consult the Maltese about the desired changes in the Constitution and there was much debating on the subject. By April 1921, Malta at last achieved responsible government. Under a new Constitution, she would have a Legislative Assembly composed of 32 elected members, and an upper house or Senate of 16 members of mixed representatives from the clergy, Chamber of Commerce, Trade Unions, University graduates and the Nobility and from two electoral districts. Self-Government had at last arrived in Malta. All internal domestic affairs would be in the hands of the Maltese. Britain would retain responsibility for defence and foreign affairs.

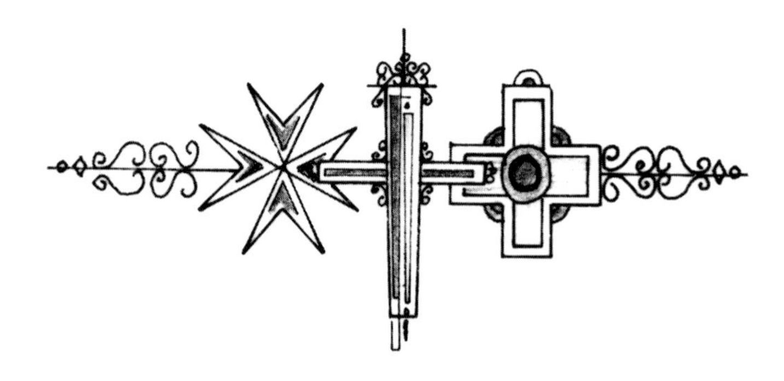

The new Constitution brought about the consolidation of Maltese political parties, as we know them today. Soon, they formalized themselves and become institutions in the political life of the country. Out of Panzavecchia's grouping was formed the 'Unione Politica'. Enrico Mizzi, still as fiery and pro-Italian as ever, emerged as the leader of the 'Partito Nazionalista Democratico'. Sir Gerald Strickland, having returned to Malta in 1920, re-entered the political arena with the 'Anglo-Maltese Party' dedicated to the supremacy of English over Italian, the encouragement of the Maltese language and to criticism of all Italianate factions including the 'unprogressive' elements within the Church. Another group dedicated to the advancement of the English language, led by Augustus Bartolo, called themselves the 'Constitutional Party'.

It was about this time that the newly formed Labour Party emerged as a political unit. It had largely grown out of the burgeoning trade union movement which had received much inspiration and support from English employees in the dockyards just after the First World War, and later, from the Workers Union in England. However, the strict origins of the labour movement in Malta can be attributed to the activities of the 'Society of Workers' which was formed by a group of workers in Senglea before the turn of the century. Being in favour of the dissemination of the English language, they were accused of being pro-Protestant by the Mizzi faction. To counter this charge they had re-named themselves as the 'Society of Catholic Workers'. The fiery and combative radical, Manwel Dimech, took up the cause of the society and, by lambasting both the British Government and the established Church, rather isolated himself and his group from the mainstream of society in Malta. The cause was then taken up by the trade union movement which was then evolving. In 1920, a 'Camera del Lavoro' was established. Basing its programme on 'Rerum Novarum', the encyclical of Pope Leo III on the condition of the working classes, it drew much support. In 1921, it was sound enough to give birth to the Labour Party, under the leadership of Savona's son, William. Highly organised, it was dedicated to compulsory education, the promotion of both the English and Maltese languages, increased direct taxation, abolition of the bread tax, parity between Maltese and British personnel in the Armed Forces and a host of other improvements in social services.

Soon, Strickland and Bartolo united under the Constitutional Party. In the election that followed the 'Unione Politica' under Panzavecchia emerged victorious with 14 seats. Labour and the Constitutionalists achieved seven seats each and the Nationalists under Mizzi, four seats. Panzavecchia declined the leadership of the Government and his deputy, Joseph Howard, became Prime Minister. The Maltese Parliament was opened by the Prince of Wales, later Edward VIII, in November 1921.

The early spirit of conciliation between the parties and the Government was certainly remarkable, although it was clear from the start that Strickland would prove to be unruly on the floor of the House, provocative in his use of language against his opponents and a disquieting influence on the political scene. The language question with its usual implications for religion again became a burning issue. In the election of 1924, Strickland was successful in winning as many seats as the 'Unione' which now turned to Mizzi's party for support. Thus both parties united under the name of the 'Nationalist Party'. Sir Ugo Mifsud led the Government as Prime Minister till the next election of 1927. The contest was now bitter and the uncompromising attitudes of the leaders of the parties on the issues of language and religion did not augur well for the future. With Strickland clearly emerging as a major force, he was dubbed as anti-clerical and pro-British. Still, with the tacit support of the reduced labour seats he managed to win a clear majority in Parliament.

As Prime Minister, Strickland lost none of his vigour and combative spirit. In the main, he was at odds with the Senate where, because of complications in the nomination of the trade union representatives and the presence of those of the hostile clergy, support for his programme foundered. He now started to shower a torrent of abuse against the clergy which, he felt, was clearly impeding progress. Soon, the Church declared war on him. He was accused of trying to make the Church subservient to the State. The British Government quickly sought to patch up the matter with the Vatican; but Rome held its ground. Strickland was now branded as an enemy of the Church. In 1930, with the election campaign in full swing, the Archbishop issued a pastoral letter exhorting all Catholics in Malta not to vote for Strickland and his followers. As a result, the British Government suspended the election. The Constitution was withdrawn; but ministers were retained in a consultative capacity. Immediately, it was claimed that Britain's action was unconstitutional. The debate over the extent of British responsibility under the diarchy of power in Malta thus commenced.

It would take two years for the dust to settle and for a new Constitution to come into effect. However, in the intervening period, the tuition of Italian in elementary schools was withdrawn and the use of the Maltese language was extended. This was violently opposed by the Nationalist party. Strickland's earlier dispute with the Church now caused him to lose seats in the next legislature. The Nationalists, clearly invigorated, fought a spirited campaign; almost a 'holy' one, judging by their emphasis on the 'Protestants under the bed' issue which underlined their determination to restore the Italian language as a means of safeguarding the Faith. Dr Paul Boffa of the Labour Party was also accused of anti-clericalism. The Nationalists thus won a resounding victory in the election of 1932 and Sir Ugo Mifsud again became Prime Minister. Appropriately, in line with party thinking, he appointed Enrico Mizzi as Minister of Education.

Mizzi's political leanings towards Italy were hardly ever disguised in his persuasive rhetoric. He immediately re-introduced the Italian language in the elementary schools, created a boy's organization called 'balilla', which was not dissimilar to that in Fascist Italy and generally encouraged the infiltration of Italian influence. This was clearly of concern to the British, who were already most apprehensive of Mussolini's ambitions in the Mediterranean, which was already being refered to as 'Mare Nostrum' in Fascist jargon. Sooner or later, the impetuous Mizzi would over-step his mark. His enthusiasm to make up for lost time in strengthening the base of the Italian language in Malta led him to include it as a major item in the list for supplementary expenditure. Regrettably, its inclusion would bankrupt Malta's reserves. For this reason, the Governor, General Sir David Campbell, had occasion to declare a state of emergency and suspend the Constitution in November 1933. The Nationalist administration had been short-lived. Malta again reverted to a colonial administration and direct rule.

The immediate response of the British Government was to end once and for all, the language dispute. In one stroke, the use of Italian in the law courts was eliminated and the Maltese language replaced it. Maltese and English became the joint official languages of the Islands. Soon, a purge was made on all Italian institutions in Malta. For six years, the largely autocratic rule of the Governor held sway. With Maltese politicians clearly agitating for a return to self-government and dominion status for Malta, the Constitution carried Malta into the war years and until the restoration of self-government in 1947. Six of the ten elected seats on the new Council of Government were won by Strickland. The Nationalists had three seats and Labour obtained one. Sitting in that Council were three future Prime Ministers: Dr Paul Boffa representing Labour and Dr Enrico Mizzi and Dr Giorgio Borg Olivier on the Nationalist side. In 1945, Mr Dom Mintoff, another Prime Minister in the making also made his first appearance on the elected bench on the side of Labour.

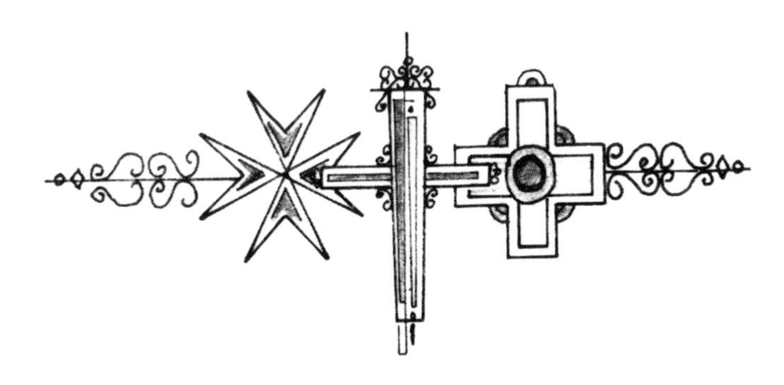

WORLD WAR II

THE SECOND SIEGE OF MALTA

By the middle of the Nineteen-Thirties, it was clear that the democracies of Britain and France were on a collision course with the totalitarianism of Germany and Italy. The German re-occupation of the de-militarized Rhineland in March 1936, followed by the signing of the military pact between Berlin and Rome in November of that year, were clear indications of the ambitions of Nazi Germany and Fascist Italy. The feeble 'policy of appeasement' adopted by the British and the French was itself contributing to the increasing militancy of the dictators in Europe. Clearly, it was doomed to failure. Sooner or later, the expected blow-up would have to come. War in Europe was inevitable.

Malta had long been a pivot in the strategy for the defence of British interests in the Mediterranean. Despite the near-certainty of war in Europe, Britain was slow to re-arm. Malta's defences were largely neglected in the pre-war years. Her role within the modern concept of war, in which air power would be a telling factor, was still most unclear. Malta was singularly ill-equipped and inadequately armed to deter any potential aggressor. Her anti-aircraft defences were minimal, air cover in the form of modern fighter aircraft was non-existent and her coastal defences were primitive and out-dated. Moreover, her reserves of food, supplies and other essential commodities were meagre and most inadequate to sustain her through an indefinite period of siege and blockade.

With the commencement of the war with Germany in 1939, it was immediately clear that, should Italy also declare war against Britain and France, Malta would be totally isolated in the central Mediterranean. For supplies and support, she would have to rely on Gibraltar, 1000 miles to the west and on Alexandria, and equal distance to the east. Should France be overwhelmed. the western Mediterranean would be seriously jeopardized. French bases in North Africa would be neutralized. Malta's isolation would then be complete. She would be at the mercy of the Italians, thirty minutes away in Sicily.

All these events would come to pass. Mussolini declared war against Britain and France on 10th June, 1940. The next morning, Italian bombers quickly attacked Malta. The first casualties were six Maltese gunners of the Royal Malta Artillery, killed outright by a high explosive bomb as they manned their guns at Fort St Elmo. With that first bomb, Malta came face to face with reality. Italy was now her declared enemy. With that first bomb too, the Italians lost all their credibility and former goodwill amongst the Maltese. The Islanders now rallied to the Allied cause, promptly gathered their resources of fortitude and courage and immediately prepared themselves for a long and painful siege.

Much would have to be done, since Malta was clearly defenceless at this stage. The immediate priority was to provide shelter for the civilian population. A gigantic programme to excavate underground shelters in all towns and villages was quickly mounted. The old railway tunnels and historic catacombs were soon converted for this purpose. With the help of experience miners from South Wales and Yorkshire, serving with the Royal Engineers in Malta, the authorities were successful in providing adequate protection for the population within a year. The early completion of this crash 'building' programme greatly contributed to the relatively low figure of civilian casualties registered in Malta during the war. However, with some foresight, more lives would have been saved.

The Italians continued their bombing raids over Malta. Initially, the 200 odd aircraft of the 'Regia Aeronautica' were only opposed by four Gloster Gladiators. Soon depleted to three, the Glosters put up a spirited fight in the air. Nicknamed Faith, Hope and Charity, they battled alone, day and night for three weeks. But soon, four Hurricanes were at last diverted to Malta to join them in their ordeal. By the end of July, a further dozen Hurricanes arrived. Against this stronger opposition, the Italian raids became noticeably less effective. Italy thus failed to subdue Malta in time. This would ultimately be most detrimental to the Axis effort in North Africa.

It was becoming increasingly clear that the Battle for Malta was also the battle for control over the 'waist' of the central Mediterranean. With the build-up of Axis forces in North Africa, their supply lines from Sicily were assuming more importance. The singular failure of the Italians to silence Malta and effectively blockade her supply lines, despite little or no opposition, proved to be of great concern to the German High Command. Clearly, Malta-based aircraft, shipping and submarines had to be prevented from ever taking to the offensive since the Axis lifeline from Sicily to North Africa would otherwise be jeopardized. For this reason, it was decided that the German Luftwaffe should move in, take over from the Italians and 'finish' the job in Malta once and for all.

With the Germans installed in Sicilian airfields by December 1940, the siege of Malta commenced in earnest. Soon, they gained air supremacy over most of the Mediterranean. Their bombing raids over Malta intensified. Mercilessly, German dive-bombers now attacked Malta's installations, dockyards and airfields. The 'box barrages' of the Maltese artillery would not deter them. The high-level aerial bombardment techniques which the Italians had previously adopted were immediately discarded. The Germans prefered to swoop down on to their targets. Their brave tactics yielded impressive results. In January, 1941, after a sustained attack on the British Fleet, as it escorted a convoy on its way to Malta and Greece, the Luftwaffe badly damaged the aircraft carrier, HMS Illustrious. Seriously on fire and crippled, she just about managed to make her way to Malta for repairs. But the Germans soon struck again. The next day, over 70 dive-bombers appeared over Malta. The Illustrious bore the brunt of their fury as she lay in dock. The surrounding areas of the dockyards and the Three Cities were also badly mauled.

During the month of January, 1941, the Germans made 57 raids on Malta. With Rommel now preparing to redress Italian reverses in North Africa, German raids on Malta were intensified. This greatly assisted the Afrika Korps to win control of Cyrenaica. In no time, Rommel was knocking on the gates of Egypt. By the spring of 1941, Greece and Crete also fell to the Germans. These gains now posed a serious threat to Malta's supply line from Alexandria. German strategy to strangle Malta to submission was clearly bearing fruit. But suddenly, a reversal of German policy saved Malta.

In June 1941, Hitler committed his all important blunder by attacking Russia. He was thus compelled to substantially reduce the German strike force in Sicily and redeploy it to that new front. In the circumstances, Malta recieved a temporary respite and, for a while, raids became fewer. At last too, the British were now awakened to the importance of Malta's strategic role in the battle for the control of supply lines in the Mediterranean. Allied reverses in North Africa had been largely caused by Rommel's ability to keep his supply links with Italy and Sicily secure. That relative security had depended on the Luftwaffe's air superiority over Malta. As this started to fade, the British promptly took to the offensive from Malta for the first time. With the arrival of reinforcements, Malta was soon in a position to fulfil her new defined strategic role. The Axis powers in North Africa would be blockaded and deprived of their supplies. From Malta, the British now launched their offensive against Axis shipping. Blenheim aircraft and Wellington bombers commenced their raids on Naples and other Italian ports. Hurricanes and Beaufighters systematically attacked targets in Sicily and Sardinia. Tripoli in Libya, also came in for a heavy pounding. At sea, Malta-based submarines preyed on Axis shipping with remarkable success. The combined results of this fresh Allied offensive in the Mediterranean brought Rommel's progress in North Africa to a halt.

It was about this time that the Italian Navy, inspired by its recent success against British shipping in Crete, made a bold attempt to breach the defences of the Grand Harbour and attack the submarine base and a recently arrived convoy. It was a singularly brave bid which, although ending in total disaster, aroused the admiration of Maltese and British alike for its daring, and for the courage and heroism of the Italians. The plan had called for two 'manned' torpedoes to blast an entry through the steel netting of a gap in the breakwater at the entrance to the Harbour. Coming up from the rear, nine explosive boats were then to career through the opening, enter the Grand Harbour and finally home in to their selected targets. In the early hours of the appointed day for the attack - 26th July, 1941 - the Italians approached Malta from Sicily. However, well before the assault craft commenced their final perilous route in the direction of the Grand Harbour, the escorting flotilla was detected by Maltese radar and the coastal defences were immediately alerted. The timing for the explosion of the 'human' torpedoes, once they reached their target, did not go off quite as planned. The outcome was that the Italian crews bravely persevered on their collision course till the moment of impact with the netting and thus blew themselves up with the explosion. But the damage caused to the steel net was minimal. Coming up from the rear, the leading E-boat, equally bravely, pursued another collision course in the hope of breaching the obstruction and thus blast an entry for the other E-boats. However, the next explosion only caused the iron bridge which lay over the netting to collapse and further obstruct the desired opening. Immediately, the coastal searchlights now came on and the guns of the Royal Malta Artillery at Forts St Elmo and Ricasoli opened fire on the stranded E-boats. Within six minutes, they were all wiped out.

With the Allies making sweeping gains in North Africa in the autumn of 1941, the Germans came to realise their earlier mistake in allowing Malta to take to the offensive when their fighter aircraft were withdrawn from the Sicilian airfields. It was now clear that Malta was the chief menace and she had to be neutralised once and for all. By December 1941, the Luftwaffe was restored in strength in Sicily and the bombing of Malta re-commenced with a vengeance. At the same time, plans were also laid for a German invasion.

Because of Rommel's desperate need for supplies to get through to him in North Africa, the Luftwaffe now concentrated all its resources on Malta. The Island's offensive capacity had to be subdued. During the month of December, the Germans subjected Malta to intensive bombardment with a total of 169 raids. Their success was immediately apparent since supplies to Rommel now started to get through and by January, 1942 he would again be in a position to stage a brilliant come-back and re-conquer Cyrenaica. The tables were thus turned again. Again, Malta's supply route from Alexandria was in jeopardy. Again, Malta was isolated and on the defensive.

From now on, the defence of Malta would be crucial to the Allies. Time and again it had been proved that Malta was the key to success or failure in North Africa. In a message to Malta, Winston Churchill tried to raise the Island's morale. "The eyes of all Britain and, indeed, of the British Empire are watching Malta in her struggle day by day, and we are sure that her success as well as glory will reward your efforts". That success was very much in doubt in the early months of 1942 since, day by day, Malta was being pounded and blitzed incessantly by the Luftwaffe. In January, the Germans carried out 263 raids over Malta. In February, 1000 tons of bombs were dropped. March saw a further intensification of the bombing raids. By April, a peak of 283 bombing raids was reached and a total of 6700 tons of bombs were unloaded onto the Island.

The situation in Malta was now precarious and even desperate. The airfields came under constant attack. There were severe losses of aircraft trapped on the ground, despite round the clock efforts to repair the holed runways. Soon, the blockade of Malta began to have a telling effect on the Island's reserves of stores, munitions and fuel. In time, the acute shortage of anti-aircraft shells for Malta's gunners would cause some guns to remain idle. With food in short supply, rations had to be drastically reduced. Government-run 'Victory Kitchens' were introduced to feed the now starving population. The necessities of life were all hard to come by. Sugar was unobtainable, edible oil became a rarity and even soap and matches had to be rationed.

As buildings collapsed everywhere, many urban areas were reduced to a heap of masonry and rubble. In April 1942 alone, more than 11,000 buildings were destroyed or damaged. Because of their proximity to the Naval Dockyards, the Three Cities were particularly badly hit. In Valletta too, many historic buildings recieved direct hits. The Royal Opera House, the Law Courts and some of the old Auberges were totally destroyed.

People were now forced to abandon their homes in the major towns and seek the comparative safety of the countryside - away from the more vulnerable urban areas. Families would spend days and nights huddled up in damp under-ground shelters. Soon, hygienic conditions in them became unbearable. The endurance of the Maltese through this harrowing ordeal was quite remarkable. The incessant bombing, the screaching of dive-bombers, the roar of the barrage, the desolation of destruction everywhere, the choking dust of crumbling masonry, the horror of violent death, the misery of the old, the consternation of the young, the panic of the very young, the pain of hunger, the confusion, the chaos and the general helplessness as convoys failed to get through, were somehow all borne with a determination to win through these uncertain times.

On the 15th April, 1942 the people's sagging morale received a welcome boost. The following message arrived from King George VI:

"To honour her brave people
I award the George Cross
to the Island Fortress of Malta
to bear witness to a heroism and devotion
that will long be famous in history".

But Malta's position was steadily deteriorating. Thankfully, because of Allied pressure on Rommel's forces in North Africa, the planned German invasion of Malta, scheduled for June 1942, had to be scrapped. Troops were diverted to strengthen the Afrika Korps, now coming to a grinding halt at El Alamein. Part of the German bomber force based on Sicily was also withdrawn. Although two supply ships, out of a total of six, managed to reach Malta in July 1942, the situation remained desperate. With barely a fortnight's supply of vital provisions and fuel left for survival in Malta, 'Operation Pedestal' was mounted on 10th August. Its specific aim was to force a convoy through to Malta. Fourteen merchant ships, with an escort of three aircraft carriers, two battleships, seven cruisers, and twenty four destroyers left Gibraltar for their important destination. For five momentous days, the convoy was flayed assunder by enemy aircraft and submarines. Soon, nine of the merchant-men were sunk, along with the aircraft carrier HMS Eagle, two cruisers and one destroyer. By the 14th, four of the remaining supply ships reached Malta. The fifth, the tanker 'Ohio' was torpedoed at sea. Completely disabled and sinking, she was lashed between two destroyers and dragged into Malta. With her precious cargo, she managed to enter the Grand Harbour on 15th August; the Feast of the Assumption. For this reason, this memorable convoy is still refered to as the 'Santa Marija convoy' to this day. Its epic story is now legend in the Islands.

The safe arrival of the 'Santa Marija convoy' in Malta marked a turning point in the Island's fortunes. Although still under siege, Malta was now in a better position to hit back. With a stronger fighting force in the air, which soon included 100 Spitfires, air superiority was achieved by October 1942. This would also coincide with Montgomery's victory over Rommel at El Alamein. With North Africa in Allied hands, the siege of Malta was finally lifted. Soon after, she became the operational launching pad for the Allied invasion of Sicily.

In June 1943, King George VI visited the Islands and received a tumultous welcome from the Maltese. As Malta basked in her glory, the final irony perhaps arrived in September of that year when, with the Italian Fleet meekly assembled in Maltese harbours, Marshall Badoglio signed Italy's final surrender document also in Malta.

In November 1943, Winston Churchill also visited Malta and saw for himself the devastation inflicted by the enemy. Three weeks later, Franklin D. Roosevelt arrived and 'in the name of the people of the United States, (1) salute the Island of Malta, its people and defenders, who in the cause of freedom and justice and decency throughout the world, have rendered valorous service far above and beyond the call of duty'.

The second siege of Malta, like the first, four centuries earlier, won the admiration of the world. Out of both sieges, Malta emerged in a state of utter devastation but totally unscathed in spirit and honour. From the last siege, Malta also emerged with the firm resolve to become at last the mistress of her own destiny.

The Air Battle Over Malta 1

Summary of Air Raids over Malta

Number of Alerts during	1940	1941	1942	1943	1944
January		57	263	25	2
February		107	236	5	–
March		105	275	7	1
April		92	282	7	1
May		98	246	30	–
June	53	68	169	30	–
July	51	73	184	10	1
August	22	30	101	9	3
September	25	31	57	4	
October	10	57	153	–	
November	32	76	30	–	
December	18	169	35	–	
Totals	211	963	2031	127	8

The 3340 alerts totalled 2357 hours and 6 minutes.

The first air raid took place on 11th June, 1940 at 6.55 a.m.
The sirens sounded the last alert on 28th August, 1944, at 8.43 p.m.
The final 'all clear' was given at 9 p.m.

Aircraft losses

1940 - 1943

Axis aircraft

Destroyed by the Royal Air Force	1252
Destroyed by Gunners	241
Probably destroyed by RAF	383
Probably destroyed by Gunners	49
Damaged by RAF	1050
Damaged by Gunners	161
Axis sorties against Malta	26000

British aircraft

Lost in the air	547
Damaged in the air	504
Destroyed on the ground	160
Damaged on the ground	231
RAF operational flying hours	112247

1. Quoted from : War Relics Exhibition Souvenir handbook
Published by the National War Museum Association, Malta
No. H2 (1979)

Malta after the war

The end of the Second World War marked the beginning of the process of de-colonization in the British Empire. Malta too, was part of that process, but her path to independence was slow and often uncertain.

A good start was made with the restoration of Self-Government in 1947. For the first time, the Labour Party won a resounding victory at the polls. Dr (later Sir) Paul Boffa became the Head of the Maltese Ministry. In 1948, his title was changed to that of Prime Minister. During the three years of the Labour administration, much progress was achieved, especially in the field of social and fiscal reform. However, the British Government's decision to dismiss workers as a result of redundancies in the Naval Dockyards, immediately caused difficulties. Boffa was unable to hold the party together. A split developed over strategy as to how best to conduct financial negotiations with an unyielding British Government, over this and other issues. As a result, Mr Dom Mintoff, an influential cabinet Minister, now emerged as leader of the Party. Boffa resigned and formed an independent Workers Party. The split largely enabled the Nationalist Party, under Dr Enrico Mizzi, successfully to contest the election in 1950. With Boffa's tacit support, Mizzi now formed a minority Government. But within three months, Mizzi died in office and Dr Giorgio Borg Olivier succeeded him as Prime Minister.

Britain's encroachment on local rights, through her control over reserved matters, as in the case of the Naval Dockyards, was resented by all the political parties in Malta. There were many heated debates over financial assistance from Britain and over Malta's future status. Borg Olivier would seek to transfer Malta from the Colonial Office to the Commonwealth Relations Office and thus acquire Dominion Status for the Islands. Mintoff would prefer to see Malta 'integrated' with Great Britain, on a basis of equality and with Maltese representatives at Westminster.

In 1951, following the collapse of the Government, Malta again went to the polls. The Nationalists were returned to power and Borg Olivier now formed a coalition Government with Boffa's Workers Party. Soon, financial difficulties again beset the Government. Negotiations with the British Government would only yield an interim agreement. In 1953, the Government again fell.

In the run-up to the election, Britain announced her rejection of the Nationalist Party's request for Malta's transfer to the Commonwealth Relations Office. Instead, she offered to transfer Malta to the Home Office. But this was unacceptable to the Nationalists. However, it did serve as an indication of the British Government's lukewarm attitude to Mintoff's ideas on integration. At the polls, the Labour Party won a slim majority over the Nationalists. However, neither party would be able to command a working majority in parliament. The Nationalist Party thus joined in a coalition with the Workers Party and a Government was formed.

Prime Minister Borg Olivier now attempted to diversify Malta's ailing economy. Like Mintoff, but perhaps not so vehemently, Borg Olivier maintained that Britain was under a clear obligation to grant adequate financial aid to Malta in compensation for the Island's help to her over the previous 150 years. In Parliament, Mintoff put up a spirited opposition to the Government. His charismatic appeal and dynamism much enhanced his stature among the electorate. His moment came in 1954. The British Government, in exercise of its rights over reserved matters, decreed that Allied forces in Malta would be exempt from taxes and customs duties. This caused uproar and much antagonism in the Island and the Government again fell.

The election campaign now saw Mintoff advancing the integration principle as a remedy for Malta's long term economic problems and as a solution to her constitutional future. The Nationalist Party strongly opposed him. Borg Olivier maintained his party's determination to continue to press for Dominion Status for Malta. In 1955, Mintoff won a clear majority at the polls and emerged as Prime Minister for the first time.

Immediately, Mintoff embarked on a series of social, educational and other reforms. At first, his proposals for integration with the United Kingdom were well received at Westminster. However, difficulties over certain aspects of the plan later arose. The Church sought guarantees for her rights and privileges in Malta and her attitude towards the proposals were hardly enthusiastic. In time, the integration plan foundered in a morass of arguments and counter-arguments. Relations between Malta and Britain became strained. With much vehemence, Mintoff maintained that Malta's 'fortress' economy over the years of British rule had distorted her proper development and even hampered her industrial growth. Much social unrest ensued when the British proposal to transfer the Naval Dockyards to a commercial firm, was made known. There were more disputes with the British Government. In 1958, financial negotiations in London broke down. In April of that year, Mintoff and his cabinet resigned with a clear indication that, henceforth, the Labour Party would fight for Malta's independence from Great Britain. The mass resignation immediately led to riots in the streets. Borg Olivier declined the Governor's invitation to form an alternative Government. The Governor restored order, suspended the Constitution and took over the Island's administration.

For four years, Malta remained in a state of limbo. The Governor, assisted by an Executive Council of non-elected appointees administered the Islands.

In 1961, after much haggling, a new Constitution was announced. It would give the Islands a measure of Self-Government. The Constitution was so designed as to create a 'State of Malta'. Britain's control over reserved matters, including foreign relations, were to be reduced, thus allowing the Maltese Government to enjoy 'enabling power' over such matters. As the Queen's representative, the Governor was to be installed as Head of State. The British Government would be represented by a Commissioner.

During the run-up to the election campaign, the already strained relations between the Church and the Labour Party deteriorated rapidiy. The election was bitterly fought. In the end, the Nationalist Party won more seats than any other party and Borg Olivier was sworn-in as Prime Minister.

Immediately, a call for independence was made by the major parties and soon, discussions got under way with the British Government. However, there was wide and serious disagreement over the terms of the proposed Independence Constitution. The Labour Party soon withdrew from the all-party Independence Conference in London. Despite continued bitter opposition, the Government pressed on with the negotiations. An agreed date for Independence had to be postponed. But soon, the Government reached agreement on a new date and the Independence Constitution, which was to be tied to a ten-year Defence and Financial accord with the United Kingdom, was finally announced. The Labour Party immediately denounced Independence under the terms negotiated by the Nationalist Government. However, on the agreed date, 21st September, 1964, Malta became a Sovereign and Independent Nation within the Commonwealth.

The immediate post-Independence years of the Nationalist administration brought much progress especially in the establishment of Malta as a tourist destination. These were, of course, the years of the run-down of British Forces. Under a development plan vigourous attempts were made to diversify the economy, ease unemployment and strengthen Malta's industrial base.

In the General Elections held in 1966, the Nationalists were again returned to power. The years following were marked by disputes which again arose with the British Government over financial aid to further diversify the economy and create new jobs in the industrial sector.

The General Elections in 1971 were again bitterly fought. But this time, the Labour Party succeeded in winning a slim majority. Dom Mintoff re-emerged as Prime Minister. In that year, Sir Anthony Mamo was also appointed as the first Maltese Governor-General. The 1964 agreements with Britain were re-negotiated after many disputes. It was finally agreed that British military facilities in Malta would be terminated on 31st March, 1979. In the intervening period, the Maltese Government would pursue policies to reconstruct the economy so as to enable Malta to do away with the need for further economic dependence on a foreign military presence. On the 13th December, 1974, Parliament enacted important changes to the Constitution and on that day, Malta was declared a Republic within the Commonwealth. Sir Anthony Mamo was also nominated as the first President of the Republic of Malta.

In 1976, the country again went to the polls and the Labour Party was returned to power with an increased majority. In that year, Dr Anton Buttigieg also succeeded Sir Anthony Mamo as President of the Republic.

The years of the Labour administration have been marked by a vigorous industrialization drive under a development plan, economic growth and significant improvements in social services. The economic and social infrastructure required for a fast rate of development has also been strengthened or newly provided. Set against the dark backdrop of a world recession and galloping inflation in many countries after 1974, the Government's achievement has been considerable.

The appointed day for the withdrawal of British military forces soon arrived. On the stroke of midnight of 31st March, 1979, the Union Jack was finally lowered. That historic moment is forever commemorated in the form of the bronze monument which now stands on the site beneath the Church of St Lawrence in Vittoriosa, in the vicinity of Fort St Angelo; the one-time hub of British naval power in the Mediterranean.

Governors of Malta

Lieutenant-General the Honourable Sir Thomas Maitland	1813 - 1824
General The Marquis of Hastings	1824 - 1826
Major-General the Honourable Sir Frederic Ponsonby	1827 - 1836
Lieutenant-General Sir Henry Bouverie	1836 - 1843
Lieutenant-General Sir Patrick Stuart	1843 - 1847
The Right Honourable Richard More O'Ferrall	1847 - 1851
Major-General Sir William Reid	1851 - 1858
Lieutenant-General Sir John Gaspard le Marchant	1858 - 1864
Lieutenant-General Sir Henry Stocks	1864 - 1867
General Sir Patrick Grant	1867 - 1872
General Sir Charles Straubenzee	1872 - 1878
General Sir Arthur Borton	1878 - 1884
General Sir Lintorn Simmons	1884 - 1888
Lieutenant-General Sir Henry Torrens	1888 - 1890
Lieutenant-General Sir Henry Smyth	1890 - 1893
General Sir Arthur Freemantle	1893 - 1899
Lieutenant-General Lord Grenfell	1899 - 1903
General Sir Mansfield Clarke, Bart.	1903 - 1907
Lieutenant-General Sir Henry Grant	1907 - 1909
General Sir Leslie Rundle	1909 - 1915
Field-Marshal Lord Methuen	1915 - 1919
Field-Marshal Viscount Plumer	1919 - 1924
General Sir Walter Congreve	1924 - 1927
General Sir John du Cane	1927 - 1931
General Sir David Campbell	1931 - 1936
General Sir Charles Bonham-Carter	1936 - 1940
Lieutenant-General Sir William Dobbie	1940 - 1942
Field-Marshal Viscount Gort	1942 - 1944
Lieutenant-General Sir Edmond Schreiber	1944 - 1946
Sir Francis Douglas	1946 - 1949
Sir Gerald Creasy	1949 - 1954
Major-General Sir Robert Laycock	1954 - 1959
Admiral Sir Guy Grantham	1959 - 1962
Sir Maurice Dorman	1962 - 1964

Governers-General

Sir Maurice Dorman	1964 - 1971
Sir Anthony Mamo	1971 - 1974

Presidents of the Republic

Sir Anthony Mamo	1974 - 1976
Dr Anton Buttigieg	1976 -

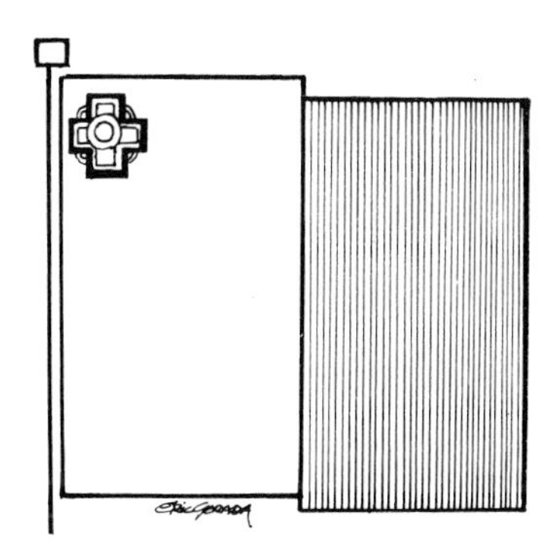

Prime Ministers of Malta

1921 - 1923	Hon. Joseph Howard, O.B.E.
1923 - 1924	Hon. Francesco Buhagiar
1924 - 1927	Hon. Sir Ugo Mifsud, K.B.
1927 - 1932	Hon. Sir Gerald Strickland, G.C.M.G. (Lord Strickland)
1932 - 1933	Hon. Sir Ugo Mifsud, K.B.
1947 - 1950	Hon. Dr (later Sir) Paul Boffa, O.B.E.
1950	Hon. Dr Enrico Mizzi, L.L.D.
1950 - 1955	Hon. Dr Giorgio Borg Olivier, L.L.D.
1955 - 1958	Hon. Dom. Mintoff, M.A. (Oxon)
1962 - 1971	Hon. Dr Giorgio Borg Olivier, L.L.D.
1971 - 1976	Hon. Dom. Mintoff, M.A. (Oxon)
1976 -	Hon. Dom. Mintoff, M.A. (Oxon)

THE PARTING SONG

Eighteen decades of history we've shared with you,
Now the hour has arrived for our sad adieu.
Not with hate in our heart
But like lovers let us part
Linked in a kiss and embrace sincere and true.

History brings about changes as times goes by ;
Which we all must bear with heads held high :
Malta today yearns to be
Without any ties and free,
Aloof from the tools of warfare - peace and goodwill nigh.-

With all her neighbours friends
And peoples of all lands,
Malta calls on all nations
To hold out their hands.

That we must reach this goal
Malta regards its role :
To strive with all its power
To labour with mind, heart and soul !

Malta's history has shown that through smiles and tears
Her hearty stock has survived hard years ;
As in the past indeed
We shall this time succeed,
Honour and glory will crown our toil and fears !

Now is the hour to part
But time will seal
Our friendship as we work
Both for the common weal.

So' ere we say goodbye
With a heavy heart,
Let us once more embrace
And kiss as we do part !

The lyrics of this poem were written on November 3 and 4, 1978 by the President of the Republic of Malta, Dr Anton Buttigieg, to music composed by Rear Admiral O.N.A. Cecil, Commander of the British Forces in Malta.

'The Parting Song' was played for the first time at the Manoel Theatre on Thursday, March 29, 1979 during a special concert by the combined bands of the Royal Marines and of the Armed Forces of Malta.

HMS Canopus in the Grand Harbour. Painted by G. Schranz 1794-1882. National Museum of Fine Arts, Valletta.

Pages 224 and 225: Monument to Sir Alexander Ball at Lower Barracca gardens, Valletta. He was the first Civil Commissioner of Malta and a popular figure among the Maltese. The monument was built out of public subscription. →

XANDRO IOAN BALL EQ. BAR.

The Royal visit to Malta by King Edward VII in 1903 - the first by a reigning British monarch.

Edward VII arrives in the Grand Harbour aboard the royal yacht 'Victoria and Albert'.

APRIL 1903

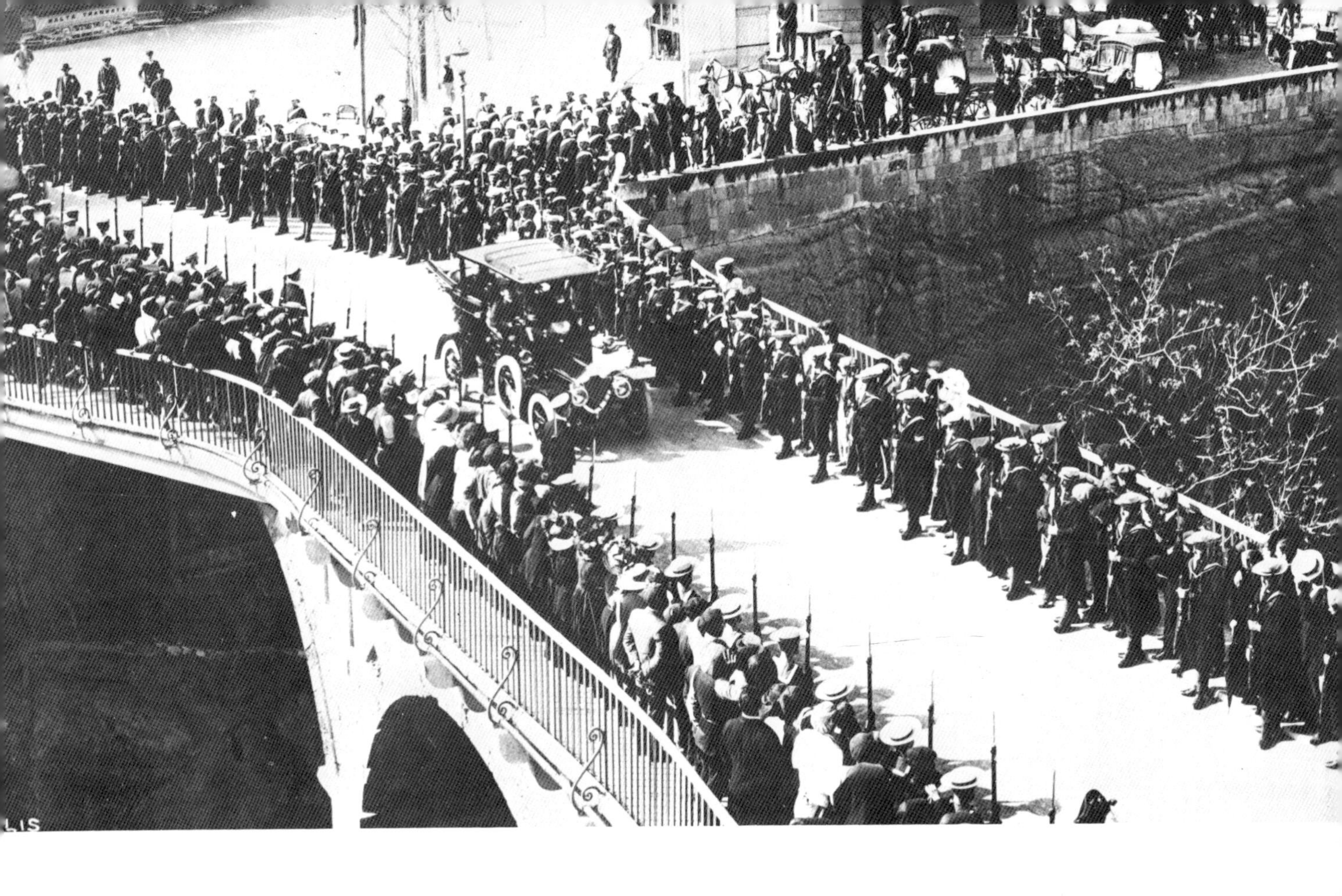

WELCOME

VALLETTA
APRIL 1903

Strada Reale, now Republic Street, decorated for the Royal visit.

The statue of Queen Victoria in festive mood for the visit of her son, King Edward VII in 19

IBLIOTHECA

King George V and Queen Mary visited Malta in 1913.

The Royal Opera House in Valletta - in its magnificence and after it recieved a direct hit from a German bomb in April, 1942.

Old Bakery Street, Valletta 1942 - a heap of rubble.

The statue of Queen Victoria in Valletta survives the bombing.

Soldiers reading the King's Message conveying the news of the award of the George Cross to the People of Malta on 11th May, 1942.

13th September, 1942.

General Lord Gort, Governor of Malta presents the George Cross to Sir George Borg, Chief Justice of Malta. A historic moment for the Islands.

The Governor
Malta

To honour her brave people
I award the George Cross to the
Island Fortress of Malta to bear
witness to a heroism and devotion
that will long be famous in history.

George R.I.

April 15th 1942

For gallantry - The George Cross.

21st September, 1964

Malta becomes a Sovereign and Independent Nation within the Commonwealth

13th December, 1974

Malta is declared a Republic within the Commonwealth

TELEPHONE

Carlo
BAR
UNITE
CLUB
YELLOW
STONE
PLAYBO

Page 236: The collonade of Mosta Church.

Page 237: Mdina gate. Erected by Grand Master Manoel de Vilhena, his armorial emblems are carved in stone over the arch.

Page 238: The Cathedral in Mdina. It was built between 1697-1702 and designed by the Maltese architect Lorenzo Gafá.

Page 239: Black cat on a balcony in Mdina.

Page 240: Blue telephone booth in a street in Floriana.

← *Page 241: Strait Street, Valletta.*

Malta Drydocks.

The Mediterranean Conference Centre in Valletta. Formerly the Sacra Infermeria or hospital of the Order of St John, it was built in 1574. Its great ward was 161 metres long - one of the longest in Europe. The Hospital became famous for the quality of its medical care. The building was badly damaged during the last war. Completely restored in 1979, it now houses the well-equipped Mediterranean Conference Centre complex which comprises six congress halls. The old great ward can be used for exhibition purposes. Republic Hall has seating capacity for 986 persons. It is equipped for simultaneous interpretation in eight languages.

lours of Air Malta - the natio

Page 246: Youth at play.

Shingle beach at Dwejra, Gozo.

Fort St Mary on the Island of Comino. Designed by Vittorio Cassar and built during the mastership of Alof de Wignacourt in the 17th Century.

TEE-H

A field of red poppies.

ST PHILIP

On the trot.

Page 252: The 'karozzin' is a horse-drawn cab. The horse takes a rest.

Page 253: Lunch break for the 'karozzin' driver.

The new Church at Manikata designed by the Maltese architect Richard England.

A golden afternoon. →

BUS
STOP

Page 260: The local handicrafts are not for the village priest.

Page 261: Mellieha village with the Church of Our Lady of Victory.

Pages 262-263: Marsaxlokk bay - a thriving fishing village. 

ONLY
Malta

End of the day.

« . . . .
But in my ears
I hear an anthem.
Is it the angels up above,
or is it
the new generation hailing our new Malta :

Our forefathers loved you, Malta our Motherland,
for centuries on end they bravely strove and fought
to win for youthat greatest honour : liberty
from all oppresive rulers and their evil thought.

Today you're independent, Malta our Motherland,
a language of your own, a mind as free,
yourname to other nations never so fair,
never again you'll sell your body at a fee.

For we Maltese do stand united in your cause,
and if needs be will spill our blood upon your altar,
so you may prosper further in progress and honesty,
and truly be a land of peace, a Maltese Malta. »

Anton Buttigieg

End-piece of his poem 'The Chapel of Peace'
Translated into English by Victor Fenech.

SOURCES AND SELECTED BIBLIOGRAPHY

History and General

Title	Author	Year
Mediterranean: Portrait of a sea	Ernle Bradford	1971
Malta: An account and an appreciation	Sir Harry Luke	1949
The Story of Malta	Brian Blouet	1967
Last Bastion	Eric Brockman	1961
Fortress, Architecture and Military History in Malta	Quentin Hughes	1969
Walls of Malta	Richard England	1973
Storja ta' Malta Vols I & II	Andrew P. Vella	1979
Malta Historical Sketches	Michael Galea	1970
More Historical Sketches	Michael Galea	1971
Malta a Shock of delight	E. Azzopardi-Sant	1932
Malta	Bernard Nantet	1975
Studies in Maltese Folklore	J. Cassar Pullicino	1976
History of the Royal Malta Artillery	A Samut-Tagliaferro	1976
The Houses of Valletta	Victor F. Denaro	1967
Why Malta Why Ghawdex	D.G. Bellanti	1964
A Maltese Anthology	A.J. Arberry	1960
Malta Isles of the Middle Sea	W. Kummerly	1965
An Outline of Maltese Geology	G. Zammit-Maempel	1977

Pre-History and Punic

Title	Author	Year
The Prehistoric Antiquities of the Maltese Islands	J. Evans	1970
Malta: An Archaeological Guide	D.H. Trump	1972
Before Civilization	Colin Renfrew	1975
Ancient Malta: A Study of its Antiquities	Harrison Lewis	1977
The Megalithic Art of the Maltese Islands	Michael Ridley	1976
The Megalithic monuments of Malta	Gerald Formosa	1975
La Missione Archeologica Italiana a Malta	S. Moscati	1963-66
Le Iscrizioni Puniche (Tas-Silg, S. Paolo Milqi)	G. Garbini	1963-66
Le Iscrizioni Puniche (Tas-Silg)	A. Guzzo	1970
Testimonianze Archeologiche a Malta	A. Azevedo	1966
The Phoenicians	D. Harden	1963
Carthage and the Carthaginians	R. Bosworth-Smith	1902
Rome and Byzantium	C. Foss & P. Magdalino	1977

Arab and Medieval

The Arabs	Anthony Nutting	1965
The Arabs	Peter Mansfield	1976
Approaches to Medieval Malta	Anthony T. Luttrell	1976
Medieval Malta: Studies on Malta before the Knights	Anthony T. Luttrell	1975

The Order of St John

The Shield and the Sword	Ernle Bradford	1972
The Order of St John of Jerusalem	Sir Hannibal Scicluna	1969
The Last Crusaders	R. Cavaliero	1960
Knights of Malta	C.E. Engel	1962
The Two Sieges of Rhodes	E. Brockman	1969
The Monks of War	D. Seward	1972
The Great Siege 1565	Ernle Bradford	1961
The Siege of Malta 1565	F. Balbi di Corregio	1566/1961
The Siege of Malta rediscovered	D. Sultana	1977
The Fortification of Malta	A. Hoppen	1979
The Shield of Europe	B. Cassar Borg Olivier	1977

French Occupation and British Period

L'Occupazione Francese di Malta	F. Cutajar	1933
Actes et Documents - Relatifs à l'histoire de l'occupation Français de Malte 1798-1800	Sir Hannibal Scicluna	1979
Britain in Malta Vols I & II	Harrison Smith	1953
Malta's Road to Independence	Edith Dobie	1967
British Malta Vols I & II	A.V. Laferla	1947
Maltese Legal History under British Rule	H.W. Harding	1968
The Malta Constitution of 1835 and its historical backround	J.J. Cremona	1959
Description of Malta and Gozo	G. Badger	1838
Party Politics in a Fortress Colony	Henry Frendo	1979
Malta: The Triumphant Years	George Hogan	1979
The Siege within the Walls	Stewart Perowne	1970
The Epic of Malta	Odhams	1943
Malta Convoy	P. Shankland	1973

Guide Books and Journals

Blue Guide Malta	Peter Mc Gregor Eadie	1979
Travellers guide to Malta	C. Kinnimouth	1967
See Malta and Gozo	Inge Severin	1978
The Maltese Islands	Charles Owen	1969
Tourist Guide to Malta and Gozo	John Manduca	1969
Holiday guide to the Maltese Islands	John Best	1975
The Year Book 1979	Hilary A C	1979
The Armed Forces of Malta Journal		April 1979
Melita Historica	Vol VII No 3	1978
Heritage - Malta's Culture and Civilization	monthly	

SAN DIMITRI POINT
ZEBBUG
Ramla bay
MARSALFORN
XAGĦRA
VICTORIA
(Rabat)
Dwejra bay
NADUR
QALA
XEWKIJA
XLENDI
MĠARR
GOZO
COMINOTTO
MARFA
MELLIEĦA
MALTA
GĦAJN TUFFIEĦA BAY
MGARR
N
SCALE
MILES
1
2
3
4
5
1
2
Kilometres
7
8
DINGLI CLIFFS

THE MALTESE ISLANDS
RICHARD PARKER
ST PAULS ISLAND
Salina bay
VICTORIA LINES
LIJA
BIRKIRKARA
ST JULIANS
SLIEMA
VALLETTA
QORMI
MARSA
THREE CITIES
PAOLA
ZABBAR
LUQA
AIRPORT
MARSASKALA BAY
ST THOMAS BAY
MARSAXLOKK
GHAR DALAM
PETERS POOL
QRENDI
HAGAR QIM
BLUE GROTTO
KALAFRANA
DELIMARA POINT

ACKNOWLEDGEMENTS

Mr. Albert Mizzi Chairman Air Malta Co. Ltd.	for making this book possible
Major Martin Scicluna	for his general collaboration in the preparation of the text and for reading the proofs
Mr. Lino Spiteri Mr. Louis Grech Mr. Charles Sacco Mr. Joe Sammut Mr. Patrick Xuereb	for their general assistance
Mr. Francis J. Mallia Director of Museums Malta Mr. Tancred Gouder Curator, Archaeological Director Malta	for their guidance
Mr. George Borg Postmaster General Malta	for use of colour separations of stamps of Malta page 3

The photographs in this book were taken by M. Christian Zuber with the following exceptions:-

Pages 226 - 230 and 231 (top)
By kind permission of R. Ellis, Valletta, Malta (Copyright)

Pages 231 (bottom), 232 (bottom), 233 (top) and 234
By kind permission the Imperial War Museum, London

Pages 232 (top) and 233 (bottom)
By kind permission The Times, Malta

Pages 184, 245 (bottom)
By kind permission Department of Information Malta

Page 244
By courtesy of Air Malta Co. Ltd.

Dr. Anton Buttigieg President of the Republic of Malta	for permission to quote from his works on pages 222 and 265

For permission from the authors and publishers to quote from works listed in the footnotes on pages:-
52 - 62 - 96 - 100 - 103 - 131 - 136 - 142 - 150 - 154 - 157 - 160 - 163 - 166 - 171 - 177 - 195 - 216

Lace photographed on page 83 is by courtesy of Carmela Cassar, Republic Street, Valletta, Malta

Items photographed on pages 114 - 192 - 225 are by courtesy of Chev. J. Galea, Mdina, Malta.

Map of the Maltese Islands and drawings on pages: 97 - 99 - 101 - 104 - 105 - 153 are by	Richard Parker
Book design and all other drawings and text by	Eric Gerada-Azzopardi

Eric Gerada-Azzopardi
London

Christian Zuber
Paris

Achevé d'imprimer sur les Presses des Editions DELROISSE
107-109-113, rue de Paris - 92100 BOULOGNE - France
Collaboration technique François DEVAUX
Dépôt légal N° 846
ISBN 2-85518-054-6

The planned citadel of Valletta. To the left is the Grand Harbour. To the right is Manoel Island flanked by Marsamxett harbour and Pietá Creek. At the extreme tip of the peninsula of Valletta lies Fort St Elmo.